AI DRIVEN NONPROFITS

A Practical Guide to Artificial Intelligence for Nonprofits

Steve MacLaughlin

Saltire Press

AI DRIVEN NONPROFITS

A Practical Guide to Artificial Intelligence for Nonprofits

by Steve MacLaughlin

Printed in the United States of America

Published by Saltire Press

July 2026: First Edition

ISBN: 0-9888507-2-9

ISBN-13: 978-0-9888507-2-9

CONTENTS

Foreword

The nonprofit sector is sleepwalking toward a cliff.

Let me be as direct as I know how. We are standing at the edge of the most consequential technological transformation in the history of fundraising (and almost everything else) — a transformation that will separate organizations capable of genuine impact from those condemned to irrelevance — and most of the sector is responding with a combination of conference panel attendance and vendor demos.

That is not a strategy. It is the appearance of engagement while the real work goes undone.

I have been building and growing nonprofits and writing about fundraising for more than half a century. I have watched every major technology wave move through this sector. Direct mail. Database systems. Online giving. Social media. Mobile. Analytics. Each time, the same drama has played out: a small cohort of organizations moved early, built genuine capability, and pulled away from the rest.

A larger cohort watched, waited, and eventually adopted — years behind, at greater cost, having ceded ground they never entirely recovered. And a stubborn tail refused, until refusing was no longer an option.

Artificial intelligence is that wave, arriving faster and with more force than anything that came before it. The question of whether to engage is already settled — AI is here, it is reshaping the environment in which every nonprofit operates, and the only real choice is whether to be shaped by it deliberately or by accident.

What is not settled — what this book addresses with the rigor and honesty the moment demands — is how to engage with it well.

We Have Been Here Before — Partially

Ten years ago, Steve MacLaughlin wrote *Data Driven Nonprofits*. I reviewed it in The Agitator the day it was published and called it a new classic. I meant it. That book changed something in our sector — not overnight, not without resistance, but genuinely. It made the case, with evidence and with stories, that the organizations doing the best work were the ones treating data not as a burden but as a strategic asset. It argued that numeracy was not the enemy of relationship fundraising. It was its most powerful ally.

Some organizations heard that message and acted on it. They built real data cultures — not just systems, but habits of evidence, disciplines of measurement, and the organizational will to let data inform decisions even when those decisions were uncomfortable. For them, the decade between *Data Driven Nonprofits* and this book was not wasted. It was preparation.

Many others did not hear it. Or heard it and nodded and went back to doing what they had always done. Their data is still messy, their decisions still run on instinct and tribal knowledge, and their donor retention rates are still a quiet catastrophe that nobody in the building talks about out loud.

I don't dwell on *Data Driven Nonprofits* here because Steve has written its successor, and the successor is more urgent. Consider the first book the foundation. What you are holding now is the building — and the building is on fire.

The Burning Platform

Here is what I know from 50 years in this work: the organizations that will thrive in the next decade are not the ones with the most sophisticated AI tools. They are the ones with the cleanest data, the strongest cultures of evidence, and the leadership willing to make the hard decisions that genuine transformation requires — before the moment of crisis forces their hand.

And here is what Steve knows, which he lays out in this book with characteristic clarity and without the comforting vagueness that passes for strategy in too much sector writing: AI is only as good as the data it learns from. Feed it dirty data — duplicate records, outdated giving histories, missing constituent information, tribal knowledge that was never written down — and it will produce outputs with complete, unearned confidence. The machine will not know it is wrong. It will be wrong at scale, at speed, and with an authority that makes the errors harder to detect and harder to walk back.

Think of it this way: AI is like a brilliant new gift officer or direct response veteran who arrived with extraordinary analytical abilities — the capacity to review thousands of donor files overnight, spot patterns invisible to human analysis, identify your best planned giving prospects from signals buried in a decade of engagement data. Now imagine that gift officer's or digital expert's entire briefing book was assembled from your uncleaned CRM. Wrong addresses. Merged duplicates. Gift histories attributed to the wrong constituents. Engagement data that was never entered because nobody thought it mattered at the time.

Those fundraisers — however talented — are going to make expensive mistakes. And they are going to make them fast.

This is not a hypothetical. It is happening right now, in organizations across the sector that have rushed to deploy AI tools on data foundations they never properly built. The results have ranged from embarrassing to genuinely harmful — not because AI doesn't work, but because the prerequisite work was skipped in the excitement of keeping up.

Steve calls this the "Nonprofit AI Paradox": AI has the potential to democratize insight and capability across the sector, but it is actually creating a new divide. The organizations with clean data, analytical cultures, and leadership capable of asking hard questions of AI-generated outputs are pulling away from those without. And the gap is not growing at a linear rate.

It is compounding — because AI capabilities build on each other, because organizations that adopt it well learn faster than those that don't, and because competitive advantages built on AI are self-reinforcing in ways that traditional operational advantages never were.

The organizations on the wrong side of that divide will not vanish. They will gradually become less effective, less relevant, less sustainable. They will find themselves — as I have watched organizations find themselves before, in every previous technology wave — explaining, years later, why they waited. The explanation will not be satisfying.

The Answer Is in Your Hands

Here, in this book, is the antidote to that outcome. And unlike some of the breathless AI literature flooding the sector, *AI Driven Nonprofits* does not promise miracles. It does not name specific vendors, platforms, or products — a deliberate and wise choice, given that the tools are changing faster than any publication cycle can track.

What it provides instead is something far more durable: a framework for thinking clearly about AI in a moment when clear thinking is in short supply.

This book is structured to serve every level of nonprofit leadership. Senior development professionals will find the most important reading in the first eight chapters, which address culture, organizational readiness, and the external forces reshaping the sector.

Frontline fundraisers will find immediately actionable material in the chapters on fundraising, programs, finance, and communication. Executive directors and board members will find in the governance and measurement chapters the frameworks for oversight that responsible AI deployment requires. And smaller organizations — the executive directors whose donor records still live in spreadsheets, the solo fundraisers who inherited databases nobody has cleaned in years — will find a section written specifically for them, meeting them where they actually are.

None of this is theoretical. Steve builds every argument on evidence from organizations doing this work right now, at every scale, across every cause area. The successes are real. The failures are instructive. And the path forward — starting with a clear-eyed assessment of your data, identifying one specific problem AI could address, building in human review loops, and measuring outcomes rather than activity — is laid out with the practical specificity that only someone who has watched these transitions up close can provide.

The Charge

Steve opens this book with a quote from Buckminster Fuller: "There is nothing in a caterpillar that tells you it's going to be a butterfly." He uses it to describe the transformation underway in the nonprofit sector — a change happening at a level many organizations cannot yet perceive, using capabilities they do not yet have. It is an honest and generous metaphor for where most of the sector sits: not failing, but in the difficult middle of becoming something it cannot quite see from the inside.

I want to add a less gentle observation to that metaphor. The caterpillar does not choose to become a butterfly. It is driven by forces beyond its control, shaped by a biological imperative that does not consult its preferences. The nonprofit sector does not have that luxury. It can choose — is choosing, right now, with every delay and every conference session and every pilot program that never becomes standard practice — what kind of organization it intends to become.

The organizations that grasp this book's argument and act on it will not look back and identify a single moment when everything changed. They will look back and recognize a series of decisions — about data, about culture, about measurement, about people — that compounded quietly until the organization they had become was no longer recognizable as the one they started from. That is not a warning. It is the most honest description available of what genuine transformation looks like.

The organizations that don't act will look back too. What they will see is equally predictable.

Read this book. Take its arguments seriously. Start where you are. Do the data work. Build the culture. And understand — as Steve understands, and as our sector must urgently come to understand — that the time available for preparation is not unlimited.

The future is being written. This book is an essential guide to writing it well.

Roger Craver
Editor, The Agitator
Author, *Retention Fundraising: The Art and Science of Keeping Your Donors for Life*

Fast Forward

The book you are about to read is filled with advice on how to adopt AI in service of your mission. Steve's last words are, "The choice is yours." So I'll pick up there and fast forward a decade to two versions of the future: one where you chose well, and one where you didn't.

Today, AI is being built fast by big companies to make a profit. They set the pace and it creates the narrative that we must keep up with their breakneck speed of development. The companies are on a different path, moving at the speed of profit.

The nonprofit sector must move more intentionally. The nonprofits that adopt AI at the speed of trust are the ones that will make the right choices that Steve has laid out.

If we make the right decisions about AI today, our future is more optimistic. By 2035, the organizations that thrived did the unglamorous work first, the work that Steve's whole book describes. They cleaned the data, measured outcomes instead of activity, and built the readiness Steve recommends.

But they also went beyond this strong foundation. They let AI in thoughtfully, and were always accountable for what went out the door. When AI handed back time, they reinvested it in their people. The grant report that ate 12 hours took 2, and the 10 hours went to a program officer sitting with a donor, not to reducing headcount.

The fundraising team killed the guilt-inducing subject lines or the thank-you note no one ever read because their data showed it wasn't working. They kept the relationships Steve says can't be automated, and they gave transactional behavior with donors the joyful funeral it deserves.

They did not simply adopt a new tool, they reinvented how they work. This created abundance — more donors, more resources, more impact.

Now the other 2035, created by different choices.

These nonprofits ran headfirst into AI to save time and fill the donor pipeline, then spent every saved hour on more tasks to get more donors in the pipeline. Steve calls it the efficiency trap. The to-do list didn't shrink. It multiplied into notifications, beeps, and constant over-working. The Ferrari had fuel, but the AI efficiency-driven fundraisers ran over their donor relationships with it.

These fundraisers only know their donors by probability scores, not their names. Their mornings come pre-sorted by a dashboard that tells them which leads to push, which synthetic thank-you to send, which urgency narrative to include in their appeals.

These fundraisers don't meet with their donors in-person or call them as their job performances are measured on pipeline velocity. Every night they go home feeling empty and a lack of purpose because they no longer feel connected to the work.

And the donors felt it before they could name it. The wealthy ones got machine-made intimacy. The smaller ones got ignored. Privacy leaked away one inferred data point at a time.

The cause turned into a glossy billboard nobody believed or trusted. And nonprofits were no longer about mission, community, and impact - they were too busy selling synthetic compassion.

Both futures ran on the same AI models, the same data. So why two different drastically different futures? Different choices made today. A thousand small choices, made one way or the other. I'd like to reinforce two important choices Steve touches in his book.

The first is whether you let people keep their own minds and not put their brains in a jar when using AI. Steve names the human skills that matter, the judgment to question an output and not push the real thinking onto whoever opens the document next.

What he doesn't dwell on is how fast a badly run AI rollout eats that judgment. Train people to hand their thinking to the tool, to take the first draft, to paste without reading, and you downskill them.

The second is who the algorithm decides to ignore. A model trained on your past major donors learns to go find more people exactly like them. So it quietly steers you toward who gets courted and who gets ignored. It scores the passionate community organizer with no money low enough to vanish. It writes off the volunteer who would have stayed 30 years.

The organizations who did that unglamorous work caught this, because they'd put the people they actually serve in the room when they chose how to use these tools, not after. Steve writes about dignity. This is what protecting it looks like, and it doesn't happen without intention.

Steve’s book gives you the foundation: the data, the principles, and the measurement. You need to decide to use the tools to carry deeper relationships and a human-centered way of working with AI.

That's the only difference between the two 2035s I mentioned earlier. Same Ferrari. Same full tank. The question was what were you driving toward.

Steve opens this book with the caterpillar that can't picture the butterfly, and it might be the most telling line he writes, because you won't feel 2035 arriving.

You'll make decisions that each seem reasonable on the day you make them. Save time as the end goal or reinvest in what matters. Train people to think or train them to let AI do their thinking. Automate relationships and purpose out of the work or keep them in. Let the model pick the donors or check who it left out.

The choice is still yours. So choose well.

Beth Kanter
Trainer, Facilitator, and Nonprofit Innovator
Co-Author, *The Smart Nonprofit: Staying Human-Centered in An Automated World*

Introduction

"There is nothing in a caterpillar that tells you it's going to be a butterfly." — Buckminster Fuller

The nonprofit sector is not short on technology conversation. Conferences have featured AI sessions for the past several years. Vendors have rebranded products, added AI to their marketing, and filled demonstration slots with capability showcases. Articles about artificial intelligence and philanthropy appear with increasing frequency in every sector publication.

Most of that conversation is happening at the wrong level. It is focused on the tools — which platform, which feature, which vendor is leading the market. That conversation is understandable and largely useless, because the tools are changing faster than any organization can track, and the organizations that focus on tools rather than principles will find themselves perpetually chasing a market that never arrives at a stable destination.

The more important conversation is about what artificial intelligence actually requires from the organizations that want to use it well — and what it makes possible for the organizations that are ready. That conversation requires honesty about data quality, cultural readiness, and the gap between what the sector is currently measuring and what it needs to measure. It requires a clear-eyed account of where AI is already producing results in nonprofit fundraising, programs, and operations — and where the expectations have significantly outrun the evidence.

This book is an attempt to have that conversation directly and practically, without the hype that has surrounded AI in the sector and without the reflexive skepticism that treats any enthusiasm for new tools as naïve. The sector deserves better than both.

The Message

The central message of *AI Driven Nonprofits* can be stated simply: artificial intelligence is a tool in service of mission — not a strategy, not a substitute for judgment, and not a solution to problems that have their roots in organizational culture and data quality.

This message has three parts.

The first is foundational. AI is only as good as the data it learns from. The organizations that will benefit most from AI are the ones that did the hard, unglamorous work of data quality before AI arrived — that built genuine measurement cultures, maintained constituent data with rigor, and developed the habit of using evidence to inform decisions. Those organizations are not starting over. They are ready. The organizations that skipped the data work and are hoping AI will compensate for what they haven't built will be disappointed in ways that are expensive and entirely predictable.

The second is practical. AI is already producing measurable results in nonprofit fundraising, program delivery, and organizational operations. Predictive models are surfacing major gift prospects that human analysis would have missed. Retention models are flagging lapsing donors before they disappear. AI-assisted communications are enabling genuine personalization at scales that were previously impossible. Impact reporting is reaching donors who have always deserved more personal attention but never received it because the economics didn't support it. These applications are real, accessible, and producing results in organizations of recognizable size doing recognizable work.

The third is strategic. The nonprofit sector is facing a set of external forces that make the internal AI conversation more urgent than most organizations recognize. Donors are changing — their expectations shaped by AI-assisted consumer experiences that now extend naturally into their philanthropic behavior. Funders are changing — foundations and government agencies are deploying AI in grant-making and compliance monitoring in ways that advantage organizations with rigorous data practices and disadvantage those without them.

And the sector is standing at the edge of a collective intelligence opportunity it has never had before — the possibility of shared learning infrastructure that could finally allow the sector to learn from itself at the speed the problems demand.

None of those forces are waiting for organizations to be ready. They are already moving.

Who This Book Is For

The primary audience for *AI Driven Nonprofits* is the people who lead and staff nonprofit development operations — chief development officers, major gift officers, annual fund directors, planned giving professionals, development operations teams, and the executives and board members who set the strategic context in which those professionals work.

But the book is written for several distinct readers within that audience, and it is worth naming them directly.

Senior development professionals and chief development officers will find the most important reading in the first eight chapters, which address culture, organizational readiness, and the external forces reshaping the sector. These are the chapters about the decisions that only you can make — about investment, about culture, and about what AI readiness actually requires before any tool is deployed. The governance chapter addresses your accountability as a decision-maker in ways that should be taken seriously.

Frontline fundraisers — major gifts, mid-level, annual fund, planned giving — will find the most immediately actionable material in the NOW chapter and the five applied chapters covering fundraising, programs, finance, revenue, and communication. These chapters address the work you are doing today and show specifically how AI is changing the economics of that work at every giving level.

Nonprofit executive directors and CEOs should prioritize the CULTURE, FORCES, and PEOPLE chapters. They address the leadership decisions — about organizational culture, external positioning, and staff capacity — that only you can make. AI adoption is ultimately a leadership question before it is a technology question.

Board members will find the GOVERNANCE and MEASURE chapters most directly relevant. These establish what responsible oversight of AI deployment looks like and what accountability frameworks should be in place before significant AI investment is made. If your organization's leadership is asking the board to approve AI-related expenditure, those two chapters provide the framework for asking the right questions.

Smaller organizations starting from scratch will find a section in the START chapter called "Starting from Nothing" written specifically for you — for the executive director whose donor records live in a spreadsheet, the solo fundraiser who inherited a database nobody has cleaned in years, the development director who has never tracked first-year retention. The path is the same path every organization follows. The starting point is different. That section addresses where you actually are.

Consultants, technology providers, and sector intermediaries will find the argument most relevant to their work in the FORCES and FUTURE chapters — specifically the case for sector-level infrastructure and shared learning capability that the market has not yet built.

A Note on the Tools

This book does not name specific AI vendors, platforms, or products. That omission is intentional.

The tools available to nonprofit organizations are changing faster than any publication cycle can track. Platforms that were category leaders at the time this book was being written have already been superseded. Startups that raised significant funding specifically to serve the nonprofit AI market have already consolidated, pivoted, or disappeared. Any vendor list published here would be partially wrong by the time most readers encounter it and confidently wrong within three years.

What this book provides instead is a framework for evaluating any tool: Does it address a specific, defined problem? Does it require data infrastructure the organization actually has? Can its results be measured against a criterion established before deployment? Those questions apply equally to whatever the market produces next year and the year after that. The tools will keep changing. The questions won't.

Caterpillars and Butterflies

The opening quote from Buckminster Fuller and the quotes that open each chapter were chosen for a reason. He was describing something that resists prediction from the inside.

The caterpillar has no frame of reference for what it is becoming. The change is happening at a level the current form cannot perceive, using capabilities the current form does not yet have. That is an uncomfortable description of where many nonprofit organizations sit right now — not because they are failing, but because the transformation underway is genuinely difficult to see clearly from within it.

This book is an attempt to provide that outside perspective — to describe what the transformation looks like, what it requires, and what becomes possible on the other side of it.

The organizations that figure this out will not look back and identify a single moment when everything changed. They will look back and recognize a series of decisions — about data, about culture, about measurement, about people — that compounded quietly until the organization they had become was no longer recognizable as the one they started from. That is not a warning. It is the most honest description available of what becoming AI-driven actually looks like in practice.

The nonprofit sector has always done its best work in the space between what is and what could be. That space has never been larger than it is right now.

Steven R. MacLaughlin

Charleston

July 2026

01 CONVERGENCE

"The question of whether a computer can think is no more interesting than the question of whether a submarine can swim." — Edsger W. Dijkstra

November 30, 2022
A Wednesday morning that would change everything. No headlines. No press conferences. Just a quiet launch of a conversational AI system called ChatGPT. Within five days, one million people had signed up. Within two months, one hundred million. The fastest technology adoption in human history.

The world woke up to artificial intelligence — not as science fiction, not as a distant possibility, but as something anyone could use right now. Something that could write, analyze, code, explain, and create. Something that felt, for the first time, genuinely intelligent.

But here's what most people missed in those early weeks: The AI revolution wasn't actually starting on November 30, 2022. It had been quietly underway for decades. ChatGPT was just the moment when the general public finally caught up to what data scientists, technologists, and a handful of forward-thinking nonprofit leaders already knew.

We had been living in the age of data driven everything. Now we were entering the age of AI driven everything. And the nonprofit sector — which had spent decades struggling to become more data driven — suddenly faced a new imperative: become AI driven or risk irrelevance.

April 5, 1905

Let's travel back 117 years before ChatGPT launched. A cool spring morning in the nation's capital brought with it nearly an inch of rain. The historical weather records tell us the sun eventually made an appearance in Washington. The temperature would reach a high of 77 degrees that Wednesday, especially warm for that time of year in the District of Columbia.

The front page of *The Washington Post* reported on President Roosevelt's speech in Louisville, Kentucky before the start of a vacation trip to the Southwest. There was a lead story about the newly elected mayor of Chicago and another headline regarding the arrest of a suspected terrorist in Saint Petersburg, Russia following a bombing in Warsaw just a week before.

On page two of *The Post* was a lengthy article headlined "MONEY IS BEST TALKER" that reported on a fundraising dinner held the night before. The Washington Young Men's Christian Association (YMCA) had attempted to raise $300,000 for a new building for the past three years. This was no trivial amount of money — equivalent to over $11 million today. Featured speakers at the dinner included Charles Ward, international secretary of the YMCA, and Lyman Pierce, general secretary of the Washington YMCA. These two men would go on to transform nonprofit fundraising, but first, there was the small matter of raising the remaining $80,000 for the building project.

Ward and Pierce did something revolutionary for their time. They brought systematic thinking to fundraising. They developed formulas. They tracked metrics. They tested approaches. They analyzed what worked. In an era without computers, databases, or even electric calculators, they pioneered data driven fundraising.

Together they developed a simple formula for fundraising success: Time-bound campaigns launched with sizable pledges already secured from major donors; the use of publicity to generate attention; and campaign clocks and thermometers to remind the public about progress along the way. These tactics would raise more than $27 million during the next decade for the YMCA alone.

Between 1905 and 1923, Ward and Pierce would help nonprofit organizations around the world raise more than $350 million — equivalent to more than $5 billion today. They did all of this without donor management systems, without analytics platforms, without email, without websites, without artificial intelligence.

They created the art of fundraising. Now, 120 years later, we're witnessing its convergence with AI. The opening chapter of *Data Driven Nonprofits* has transformed into the opening chapter of *AI Driven Nonprofits*.

The Long Arc

The story of nonprofit evolution has always been about tools converging with human purpose. Ward and Pierce used index cards, typewriters, and telegraph wires to systematize fundraising. Later generations added databases, spreadsheets, and websites. Each wave of technology promised to revolutionize the sector. Some delivered. Many didn't.

Then came the era of Big Data. Organizations accumulated vast oceans of information about donors, programs, and impact. The problem wasn't access to data — it was making sense of it all. Most nonprofits drowned in data while starving for insights.

In 2016, I wrote *Data Driven Nonprofits* to help organizations navigate this challenge. The book explored how nonprofits could transform data from overwhelming burden into strategic asset. It told stories of organizations that had cracked the code — turning information into insights, insights into decisions, and decisions into impact.

That book documented a critical evolution: from opinion-driven to data-informed decision-making. From tribal knowledge to tested hypotheses. From guesswork to measurement. It was about building cultures where data wasn't a foreign object but a natural part of daily work.

But here's what I didn't fully appreciate in 2016: We were standing at the threshold of an even more profound transformation. The data revolution was just the opening act. The AI revolution would be the main event.

Why AI Changes Everything

The fundamental difference between being data driven and being AI driven comes down to who does the cognitive work. Data driven organizations use information to inform human decision-making. Humans ask questions. Humans analyze patterns. Humans draw conclusions. Humans decide what to do next. Data is the raw material, but humans do all the cognitive work.

AI driven organizations augment human intelligence with machine intelligence. AI can ask questions humans never thought to ask, spot patterns invisible to human analysis, and generate insights at scales and speeds impossible for human cognition alone. This isn't about replacing humans. It's about amplifying human capability — freeing nonprofit professionals from the drudgery of data wrangling so they can focus on relationship building, strategic thinking, and creative problem-solving.

Consider what this looks like in practice across a single organization. A fundraiser queries the donor database in plain language — *show me donors who gave for the first time in 2023, made their second gift in 2024, and have capacity for a $1,000 annual gift but are currently giving less than $250 —* and gets back not just a list but an interpretation: why these donors match, what the giving pattern suggests, where the opportunity is. A program manager uploads months of unstructured beneficiary feedback and receives a synthesis of recurring themes, what's working, and what needs attention — context and nuance included, not just keyword counts.

A communications director asks for year-end appeal subject lines calibrated to donors who care about environmental justice and have given between $100 and $500 — and gets drafts that apply what is known about donor psychology and email performance, not just text that fills the space. A development director asks which major gift prospects deserve visits this quarter, weighted by giving history, capacity indicators, engagement patterns, and likelihood to give — and gets a ranked list with reasoning that can be challenged, refined, and acted on.

In each case, the AI is not running a query. It is interpreting intent. Data that was locked in systems becomes conversational. Analysis that required specialized skills becomes accessible. Insights that took weeks to generate become instantaneous. Tasks that consumed hours get automated. The cognitive work that used to bottleneck in human bandwidth now has a capable partner.

The Nonprofit AI Divide

But here's what concerns me: We're already seeing a dangerous divergence in the nonprofit sector.

Some organizations are moving fast. They're experimenting with AI tools. They're training staff. They're rethinking workflows. They're measuring results. They're building competitive advantages that compound daily.

Other organizations are paralyzed. Paralyzed by fear of the unknown. Paralyzed by concerns about ethics and bias. Paralyzed by budget constraints. Paralyzed by the overwhelming pace of change. Paralyzed by leadership that doesn't understand AI. Paralyzed by the same organizational inertia that kept them from becoming data driven.

The gap between these two groups is not growing linearly. It is growing exponentially — because AI capabilities compound, because organizations that adopt AI learn faster than those that don't, and because competitive advantages built on AI are self-reinforcing in ways that traditional operational advantages never were.

In 2016, I warned about the Nonprofit Data Paradox: the more data you have of something valuable, the less valuable it becomes if you can't use it. Organizations were drowning in data while starving for insights.

Now we face the Nonprofit AI Paradox: AI has the potential to democratize insight and capability, but it's actually creating a new divide. Organizations with the resources, leadership, and culture to adopt AI will pull away from those without. The AI empowered get better results. The rest get left behind.

This isn't just about competitive fundraising. It's about impact. The organizations that harness AI will be able to serve more people, more effectively, with fewer resources. They'll be able to predict needs before they become crises. They'll be able to personalize at scale. They'll be able to measure what matters. The organizations that don't will find themselves trapped in 20th-century workflows trying to solve 21st-century problems. The gap isn't just about efficiency — it's about effectiveness, relevance, and ultimately, survival.

Not Starting from Zero

Here's the good news: If your organization did the hard work of becoming more data driven over the past decade, you're not starting from scratch. The foundations you built — data quality, analytical thinking, evidence-based decision-making, measurement culture — these are exactly what you need for AI.

In fact, organizations that skipped the data revolution and try to jump straight to AI will struggle. Because AI without good data is like a Ferrari without fuel. It might look impressive, but it won't go anywhere. Bad data doesn't just limit AI — it also actively misleads it.

The journey to becoming AI driven isn't separate from becoming data driven — it's the natural evolution. It's what comes next. Data gave us eyes to see patterns. AI gives us the capacity to act on them at scale.

But — and this is critical — you don't need perfect data to start with AI. You don't need massive technology budgets. You don't need data scientists on staff. You don't need to understand how neural networks work.

You need curiosity. You need willingness to experiment. You need leadership that understands both the opportunity and the responsibility. You need a culture that embraces learning. You need champions at all levels who will test, learn, share, and iterate.

Most importantly, you need to start now. Not when you have perfect data. Not when you have more budget. Not when you fully understand AI. Now.

What This Book Will Do

AI Driven Nonprofits picks up where *Data Driven Nonprofits* left off. It assumes you understand why data matters. It builds on that foundation to show you how AI amplifies everything data promised but couldn't quite deliver.

This book will help you understand:

How AI actually works — not the technical details, but the practical realities nonprofit leaders need to grasp.

Where AI creates value — the specific use cases where AI transforms nonprofit operations, from fundraising to program delivery to impact measurement.

How to get started — practical first steps that don't require massive investment or technical expertise.

What good looks like — real examples from nonprofits already succeeding with AI.

How to navigate risks — addressing legitimate concerns about bias, privacy, transparency, and responsible AI use.

How to build AI culture — because technology without culture change is just expensive decoration.

What the future holds — where AI is heading and what it means for the nonprofit sector.
Like *Data Driven Nonprofits*, this book is built on real stories from real organizations. You'll meet nonprofit leaders who are navigating this transformation right now. You'll learn from their successes and their failures. You'll see what's actually working in the field, not just what sounds good in theory.

The Stakes
Let me be direct about why this matters.

The problems nonprofits exist to solve — poverty, disease, injustice, environmental destruction, educational inequality — are getting more complex and urgent. At the same time, competition for donor dollars is intensifying. Government funding is uncertain. Traditional approaches are producing diminishing returns.

AI isn't going to solve these challenges by itself. But organizations that harness AI will have capabilities that organizations without AI simply can't match. They'll identify at-risk populations before crises occur. They'll personalize engagement with millions of supporters. They'll measure impact with unprecedented precision. They'll be able to prove their effectiveness in ways that attract more support — creating a cycle of innovation that compounds over time.

The organizations that don't adapt won't disappear overnight. But they'll gradually become less relevant, less effective, less sustainable. They'll be stuck using yesterday's tools to fight tomorrow's battles.

This isn't fear-mongering. It's pattern recognition. We've seen this before with every major technology shift. The organizations that adapted thrived. The organizations that resisted found themselves explaining, years later, why they waited.

An Invitation

So here's my invitation: Don't wait for perfect understanding. Don't wait for perfect conditions. Don't wait for someone else to go first.

Start where you are. Use what you have. Learn as you go. This book is your guide. Not a technical manual — a practical roadmap. Not a crystal ball — a compass. Not a replacement for judgment — a tool to enhance it.

The convergence is here. Data and AI. Human purpose and machine capability. The art pioneered by Ward and Pierce meeting the science enabled by neural networks.

It's 120 years since that Washington YMCA campaign. 10 years since many nonprofits started their data journey. And right now — this moment — is when we decide whether to embrace the AI revolution or be left behind by it.

The future is being written. Let's make sure the nonprofit sector has a leading role in the story.

Turn the page. Let's get started.

02 CONNECTIONS

"The question is not whether machines will think. The question is whether men will." — Sydney J. Harris

1725. Lyon, France.
Basile Bouchon was not a computer scientist. He wasn't an engineer or a mathematician. He was the son of an organ maker — and that accident of birth may be the single most consequential fact in the history of artificial intelligence.

Watch a pipe organ work and you begin to understand what Bouchon saw every day growing up. Rotating cylinders studded with pegs and cams that physically contacted mechanical arms, triggering notes in precise sequence. The music wasn't random. It wasn't improvised. It was encoded — written into the physical structure of the instrument itself. The cylinder didn't play music. It read instructions.

In 1725, Bouchon applied that same logic to a silk loom. He replaced the cylinder with a roll of perforated paper. Each hole in the paper corresponded to a hook in the loom. Where there was a hole, the hook passed through and lifted a thread.

Where there was no hole, the hook was blocked. Row by row, the loom read the paper — and wove a pattern that had been encoded by whoever punched the holes.

It was slow. It was imperfect. It required a human operator to guide the process. But for the first time in history, a machine was following instructions that had been written down and fed into it. The logic of the loom wasn't in the machine. It was in the data.

Bouchon didn't set out to invent the foundation of computing. He was trying to solve a specific problem: how to weave complex patterns in silk without depending entirely on the skill and memory of a human draw boy. The technology that followed — Falcon's improvements, Jacquard's famous loom of 1804, Babbage's analytical engine, Hollerith's tabulating machines — all trace a direct line back to a workshop in Lyon and a young man who grew up watching an organ think.

The insight was simple. Profound things usually are: information could be stored outside the human mind. Machines could read it. Machines could act on it.
Every database, every analytics platform, every AI model running in the nonprofit sector today is downstream of that moment.

1954. Georgetown University.
IBM and Georgetown University held a press demonstration that drew packed rooms and breathless headlines. For the first time, a computer had translated Russian sentences into English — automatically, without a human translator in the loop. The machine processed sixty Russian sentences in less than a minute. The researchers predicted that machine translation would be a solved problem within three to five years.

They were off by about sixty years. But in the moment, the demonstration was remarkable.

The machine used punch cards — a direct descendant of Bouchon's perforated paper — to encode both the Russian vocabulary and the grammatical rules for translation. Feed in a card, get out a sentence. The system worked on a vocabulary of 250 words and six grammar rules. For the carefully selected demonstration sentences, it performed beautifully.
Then someone tested it on a phrase from the Bible: *The spirit is willing, but the flesh is weak.* The machine, applying its rules with perfect mechanical literalness, returned a translation that has since become legend: *The vodka is strong, but the meat is rotten.*

It wasn't wrong, exactly. Both translations were technically defensible given the words involved. But anyone who spoke Russian — anyone who understood context, connotation, and the long human history embedded in a phrase — knew instantly that the machine had missed the point entirely.

The Georgetown experiment illuminated something that would take the field of artificial intelligence decades to fully absorb: syntax is not semantics. Processing language is not the same as understanding it. A machine can follow rules perfectly and still be completely wrong.

The punch cards were more sophisticated than Bouchon's perforated paper. The machine was faster than any human translator. But the intelligence — the actual understanding of what words mean and why — wasn't in the cards. It wasn't in the machine. It had to come from somewhere else.

That problem would occupy researchers for the next seventy years. And the lesson it carried — that better tools and better data are not the same thing as better judgment — applies directly to every nonprofit deploying AI today.

1995. Redmond, Washington.

Microsoft shipped a new version of Office that included a feature the company had spent years developing. It was called Clippy — officially, the Office Assistant — and it was an animated paperclip with eyes that watched what you were typing and offered to help.

It looks like you're writing a letter. Would you like help?

Clippy was, by almost any measure, a failure. Users found it intrusive, condescending, and impossible to train. It appeared when you didn't want it. It disappeared when you needed it. Its advice was often irrelevant and occasionally wrong. By 2001, Microsoft had quietly retired it — and Clippy became shorthand for everything that could go wrong when a machine was given the job of anticipating human needs without the intelligence to actually do it.

But here is what gets lost in the mockery: Clippy was a genuine attempt to build an AI agent. Not the large language model kind we have today. Not the kind that can write a planned giving proposal or analyze a lapsed donor file. But an agent nonetheless — a system designed to observe what a user was doing, infer what they were trying to accomplish, and offer assistance without being explicitly asked.

That is, in its bones, the same thing every modern AI assistant is trying to do.

Clippy failed not because the idea was wrong. It failed because the technology wasn't ready, the data was too thin, and the implementation wasn't honest about its limitations. It promised more than it could deliver. It interrupted more than it helped. It optimized for presence rather than usefulness.

The nonprofit sector is watching the same pattern play out right now. The tools are exponentially more capable than anything Microsoft shipped in 1995. The temptation to overpromise — and the organizational pressure to appear cutting-edge before the foundation is ready — is exactly the same.

2016. Seoul, South Korea.
The match was scheduled for five games. Most observers expected it to end quickly — and it did. Just not the way they anticipated.

Lee Sedol was one of the greatest Go players alive. Go is a 2,500-year-old board game of extraordinary complexity — more possible positions on the board than there are atoms in the observable universe. Chess grandmasters had fallen to computers years earlier, but Go was supposed to be different.

The game required intuition. Pattern recognition at a depth that felt, to anyone who studied it, fundamentally human. Most experts estimated that machines were still a decade away from challenging the world's best.

DeepMind's AlphaGo won the first three games. Lee Sedol won the fourth — a moment so unexpected that observers compared it to a human achievement of historic proportion. AlphaGo won the fifth. The match was over. The decade of runway had collapsed into months.

What made AlphaGo significant wasn't just that it won. It was *how* it won. Previous AI systems had been built by encoding human knowledge — feeding the machine thousands of games played by masters, then programming it to apply the rules those masters had developed over centuries. AlphaGo had learned that way too, at first. But it had also played millions of games against itself, discovering strategies that no human had ever used — moves that expert commentators initially called mistakes before realizing, several turns later, that the machine had seen something they couldn't.

Eighteen months after AlphaGo defeated Lee Sedol, DeepMind released AlphaZero. It was given only the basic rules of Go — no human games, no encoded expert knowledge, no inherited wisdom from centuries of masters. It learned entirely from self-play. In 72 hours, it defeated AlphaGo by a wide margin.

In 2019, a researcher named Rich Sutton published a short essay that the AI field has since called "The Bitter Lesson." His argument was blunt: every time researchers built AI systems based on encoding human knowledge and human expertise, those systems were eventually beaten by simpler systems with more computation and more data. The lesson was bitter because it meant that what humans *know* is less important than what machines can *measure*. The competitive advantage isn't in accumulated wisdom. It's in accumulated data.

For nonprofits, the implication is direct and uncomfortable. Organizations that spent decades encoding institutional knowledge into processes, staff expertise, and tribal memory — but didn't invest in capturing and maintaining the underlying data — have less to build on than they think. You cannot teach a machine what you never wrote down. The giving patterns that lived in a gift officer's head and never made it into the CRM. The engagement signals that were observed but never tracked. The constituent relationships that existed in email threads and never in a record. All of it — invisible to AI. All of it — a missed foundation.

The bitter lesson for the nonprofit sector is the same one Sutton named for the research community: the organizations that will lead in an AI-driven world are not necessarily the ones with the longest institutional history. They are the ones with the best data.

2022. San Francisco, California.
No headlines. No press conferences. Just a quiet product launch on a Wednesday morning. Within five days, one million people had signed up for ChatGPT. Within two months, one hundred million. The fastest technology adoption in human history — and most of the nonprofit sector experienced it the same way the rest of the world did: through a mix of wonder, anxiety, and an immediate scramble to figure out what it meant.

Within weeks, a question was circulating at every conference, in every board room, and across every development office in the sector: *Can AI write our appeal letters?*

It was the wrong question. And the fact that it was the first question — across so many organizations, so simultaneously — revealed something important about how the sector was framing the technology. Not as an intelligence layer that could transform how organizations understand their donors, identify their best prospects, predict attrition before it happens, and surface planned giving potential that had been hiding in plain sight for years. As a content production shortcut.

The Georgetown researchers in 1954 asked: *Can a machine translate language?* They got vodka and meat — a demonstration that looked impressive until it encountered the actual complexity of the problem. Microsoft in 1995 asked: *Can software anticipate what users need?* They got Clippy — a tool that generated activity without producing value.

And in 2022, a significant portion of the nonprofit sector asked: *Can AI write our emails?* Some of them got the digital equivalent — content that looked personalized, moved quickly through production, and performed worse than what it replaced, because it was optimized for efficiency rather than relationship.

But not everyone asked the same question. The organizations that paused long enough to ask a different one — *What does AI now make possible that was impossible before?* — found themselves looking at genuinely transformative capabilities.

A mid-sized university foundation with 200,000 alumni records and no capacity to hand-score planned giving prospects suddenly had the ability to surface the top 500 with a precision that no wealth screening tool had ever approached. A regional health system using AI to flag major donor lapse risk before the donor lapsed — not after — discovered it could intervene at the moment of maximum relevance rather than the moment of minimum hope. A social services organization that had raised and spent millions without ever being able to demonstrate impact at the constituent level used AI to build outcome models that changed how it reported to funders entirely.

Same technology. Available to everyone. Completely different outcomes — determined entirely by the question asked at the beginning.

This is not a new pattern. It is the oldest pattern in the history of both technology and fundraising. Ward and Pierce had access to the same telegraph wires, the same printing presses, and the same donor populations as every other fundraiser in 1905. What separated them was how they thought about the tools. Not: *What does this machine do?* But: *What problem are we trying to solve — and can this help us solve it better?*

ChatGPT didn't start the AI era in the nonprofit sector. It started the moment when the nonprofit sector could no longer ignore it. What happens next depends, as it always has, on the quality of the question asked first.

The Thread That Connects Them

Three centuries separate Bouchon's loom from the AI platforms being piloted in development offices today. But the thread running through all of it is unbroken.

Bouchon encoded instructions into perforated paper so a machine could act without a human directing every movement.

Georgetown encoded vocabulary and grammar rules onto punch cards so a machine could produce language. Microsoft encoded behavioral triggers so a software agent could respond to context. DeepMind built systems that stopped encoding human knowledge entirely — and let the data do the teaching.

And in 2022, the world got its first genuinely accessible AI platform capable of understanding context, generating content, and reasoning across complexity at a scale no previous tool had matched.

Every generation inherited the failures and partial successes of the one before. Every generation punched new holes in the paper, added new rules to the engine, gave the machine a little more to work with.

What none of them fully solved — what we are still working on — is the gap between mechanical accuracy and genuine understanding. The loom wove the pattern without knowing what it was making. The translation machine produced sentences without grasping their meaning. Clippy recognized keystrokes without comprehending the writer's intent. AlphaZero mastered a game without ever grasping why it mattered to the humans watching. Today's large language models have closed that gap significantly — they handle nuance, infer intent, and generate content that clears the vodka-and-meat test with room to spare.

But a different version of the same problem persists, and it matters enormously for every nonprofit in this chapter's audience. The Georgetown machine failed because the rules it learned from were too simple for the complexity of actual language.

Today's AI systems fail when the data they learn from is wrong, incomplete, or unrepresentative. The mechanism has changed.

The underlying vulnerability hasn't.

This is why data health isn't a technical prerequisite that comes before the interesting work of AI. It *is* the work. An AI model trained on duplicate records, outdated giving histories, and missing constituent information will produce outputs with the same confident uselessness as a translation machine that renders spiritual metaphor as a vodka recommendation. The machine won't know it's wrong. It will deliver the result with complete conviction.

What This Means for Your Organization

The history matters not because it is interesting — though it is — but because it is instructive. Every inflection point in the development of artificial intelligence has come with the same temptation: to assume that a more powerful machine solves the fundamental problem. It never does. The fundamental problem is always the same one. What is the quality of what we are feeding into it? What do we actually understand about what comes out?

Bouchon's loom needed accurate paper to weave accurate patterns. Georgetown's translation machine needed better rules and richer vocabulary to produce accurate translations. Clippy needed a far more sophisticated model of human intent to provide useful assistance. AlphaZero needed better data — not better human coaching — to surpass every human expert on earth. ChatGPT-era tools need clean, complete, well-maintained constituent records to produce recommendations that actually serve the mission. Each generation of tools raised the ceiling of what was possible. None of them eliminated the dependency on what was put in.

The organizations that will look back in ten years and recognize themselves as genuinely AI-driven will not be the ones that moved fastest. They will be the ones that understood this and acted on it. They invested in data health before AI infrastructure. They asked what problem they were solving before selecting a tool. They approached the technology the way Pierce approached every tested method: with clear objectives, honest assessment of readiness, and the discipline to evaluate results rather than assume them.

They also kept the human being at the center of every consequential decision. Not because AI isn't capable — it increasingly is — but because the mission of a nonprofit is not to produce accurate outputs. It is to change lives. And that work requires judgment, conviction, and relationship that no machine has yet encoded onto any punch card.

Connecting the Dots

The history of artificial intelligence is not a story of sudden breakthroughs. It is a story of accumulated progress — each generation building on the last, each failure teaching something the next tool needed to know.

Basile Bouchon taught us that information could live outside the human mind and that machines could act on it. Georgetown taught us that processing and understanding are not the same thing — and that the quality of what a machine learns from determines the quality of what it produces. Clippy taught us that presence is not usefulness, and that tools deployed without honest assessment of their limitations erode the very trust they were meant to build. AlphaGo and AlphaZero taught us the bitter lesson: that accumulated human wisdom, however deep, is not a substitute for accumulated data, and that the organizations which failed to capture what they knew are starting from a deficit. And the ChatGPT moment taught us — is still teaching us — that the question asked at the beginning determines everything about what comes out at the end.

For the nonprofit sector, these are not abstract lessons from the history of computer science. They are operational guidance for decisions being made right now. Whether to invest in data health before deploying AI. Whether to ask what problem needs solving before selecting a tool. Whether to keep humans in the loop on decisions that affect donor relationships, mission delivery, and organizational reputation. Whether to evaluate AI outputs with the same rigor applied to any other organizational practice — or to accept confident-looking results at face value and discover the vodka problem too late.

Three hundred years after Basile Bouchon punched his first hole in a strip of paper, we are still learning the same lesson. Better machines require better inputs. Better outputs require better questions. And the irreplaceable element — the one no generation of AI has yet encoded — is a human being with burning conviction about the cause, asking another human being to join it.

The tools will keep improving. That part won't change.

03 CULTURE

"We shape our tools, and thereafter our tools shape us."
— Marshall McLuhan

The Decoration Problem

The invitation at the close of the opening chapter — start where you are, use what you have, learn as you go — is the right spirit. But there's a catch that every experienced nonprofit leader will recognize immediately.

Technology without culture change is just expensive decoration.

That phrase was not included casually. Because the history of technology adoption in the nonprofit sector is littered with expensive decorations. CRM systems nobody uses consistently. Analytics platforms with dashboards nobody checks. Wealth screening data pulled once and forgotten. Predictive models that produce scores nobody trusts enough to act on.

Lessons from the Past

Lyman Pierce was a pioneering fundraiser during the early 1900s. His innovations in fundraising strategy and tactics live on to this day. Later in his life, Pierce recorded these lessons in his book *How to Raise Money*. The book was completely rewritten and republished in 2026.

In *How to Raise Money*, Pierce identified eleven factors that determine a campaign's potential for success before it even begins. Competent leadership sits at the top of that list — not sophisticated methods, not abundant resources, not favorable conditions. Leadership. Culture. The human foundation on which everything else is built.

"The campaigns that instantly command attention and support," Pierce wrote, "always have one thing in common: competent leadership. When strong administration exists, other success factors fall into place with relative ease."

He was equally direct about what failure looked like. The weakest spot in many organizations, he argued, was what he called "complete staff-ization" — paying people to do everything. An organization where paid staff handle all the work loses the authenticity, peer influence, and community engagement that makes fundraising actually work. The volunteer who asks her peer for a gift is more persuasive than the most eloquent professional precisely because her ask isn't transactional. She believes in the cause. She gave herself. That cannot be purchased or automated.

Fast-forward nearly a century and substitute AI for staff-ization, and Pierce's warning is more relevant than ever. The organizations that will struggle most with AI are not those that adopt it too slowly. They are those that use it to replace the human elements that made them effective in the first place — the relationships, the judgment, the authentic conviction that a cause is worth fighting for.

Pierce also identified something he considered the single most vital factor in any great undertaking: burning conviction. "The one vital thing behind getting anything done," he wrote, "is the overwhelming, burning conviction that it must be done. That's the force that creates programs, lines people up for service, and removes financial barriers."

He put it even more simply elsewhere: *A person must burn before they can shine. Knowledge of need makes people burn.*

No AI system produces that fire. What AI can do — when deployed with intention and cultural discipline — is help more people feel its heat. It can surface the stories, identify the prospects, and personalize the communication that puts knowledge of need in front of the people who need to hear it.

But the conviction has to come first. Culture is what keeps the fire from going out.

We've Seen This Before

Every major technology shift in the nonprofit sector has followed the same arc.

First comes the announcement: this changes everything. Then comes the adoption curve — early movers who experiment, a skeptical majority that waits, and a resistant tail that digs in.

Then, a few years later, comes the honest reckoning. The technology itself was never the hard part. The culture was. We lived this with data. In 2016, when *Data Driven Nonprofits* was published, the tools were available, affordable, and increasingly accessible. CRM systems held millions of constituent records. Analytics platforms offered unprecedented insight into donor behavior. Wealth screening and predictive modeling were no longer the exclusive province of large research universities and major medical centers.

And yet most nonprofits were not using them. Not because of cost. Not because of complexity. Because the culture wasn't ready.

The organizations that cracked the code weren't necessarily the ones with the biggest budgets or the most sophisticated technology. They were the ones that built the right culture first.

A culture where data was trusted. Where evidence informed decisions. Where testing was encouraged and failure was useful. Where champions at every level modeled the behaviors that others eventually followed.

That's the lesson the sector learned — slowly — about data. It's the lesson we need to apply immediately to AI.

The Nonprofit AI Paradox Starts Here

The first chapter of this book introduced the Nonprofit AI Paradox: AI has the potential to democratize insight and capability, but it's actually creating a new divide. Organizations with the resources, leadership, and culture to adopt AI will pull away from those without. The gap is growing exponentially. Culture is where that paradox either gets solved or gets worse.

Consider what happens when AI lands in two different organizations. In the first, there's already a culture of data. Data health is maintained. Analytics inform decisions. Staff have been trained to question outputs, not just accept them. Leadership asks hard questions before deploying new tools.

When AI arrives, it accelerates what already works. The culture shapes how the tool gets used — and McLuhan's principle operates in the organization's favor.

In the second organization, data is messy. Decisions still run on instinct and tribal knowledge. Nobody has consistent habits around measurement. When AI arrives, it gets deployed quickly — because the pressure to keep up is real — but it learns from dirty data, produces outputs nobody has the analytical muscle to evaluate critically, and gets quietly deprioritized after the first round of confusing results.

McLuhan's principle operates here too, in the opposite direction. The tool shapes the culture. Confidence in flawed outputs replaces the instinct to question them.

Pierce understood this dynamic long before the word algorithm existed. He warned that lack of perspective is one of the most common weaknesses in organizational programs.

"Part of the task gets magnified until it looks like the whole thing," he wrote. "'They can't see the forest for the trees.'" That's precisely what happens when organizations mistake AI adoption for AI impact — optimizing the tool while losing sight of the mission it's supposed to serve.

The Four Leadership Factors

Lyman Pierce identified four fundamentally important factors in effective fundraising leadership — factors he believed had determined more campaign outcomes than any other consideration. They were: know your field, connect strategy to reality, build your team, and master implementation.

These four factors map directly onto what AI-ready organizational culture requires today.

Know your field. Leaders must understand the facts and conditions in their area of expertise — not just internally, but the full landscape of need they exist to address. For AI adoption, this means understanding not just that AI tools exist, but which problems in your specific organization are worth solving with them.

Most AI failures in the nonprofit sector don't begin with bad technology. They begin with leaders who don't have enough command of their own operational reality to know what AI should be doing. The organizations that use AI well are led by people who know their donor file deeply, understand where their fundraising program leaks, and can describe their most important unsolved problems with specificity.

Connect strategy to reality. Programs and policies must directly relate to known facts — not aspirational ones, not convenient ones. The equivalent failure mode in AI adoption is deploying tools in service of the strategy you wish you had rather than the one you actually have. An organization that hasn't built a functioning mid-level donor program doesn't need AI to personalize mid-level donor communications.

It needs to build the program first. Technology deployed ahead of strategy produces the expensive decoration problem. Strategy clarified first, then technology selected in service of it — that's the sequence Pierce would recognize.

Build your team. Pierce was unambiguous about this: "Anyone who can't point each year to a group of strong, competent volunteers taking on responsibilities can't count themselves as a successful leader. They may work hard and handle enormous amounts of work personally, but they're not an executive."

The AI equivalent is this: any leader who can't point to staff members genuinely capable of evaluating AI outputs critically, identifying when models are producing wrong recommendations, and understanding where human judgment is essential — that leader has not built the team AI requires.

AI doesn't reduce the need for capable people. It raises the bar for what capable people need to know.

Master implementation. The fourth factor is about developing and adapting effective methods — not just adopting them, but learning to execute them with discipline over time. This is the discipline that separates organizations that extract lasting value from AI from those that run a pilot, generate a few encouraging metrics, and quietly move on. Mastering implementation means building AI into workflows, measuring its actual impact on mission outcomes, adjusting continuously, and holding the technology to the same standard of accountability that Pierce applied to every element of a campaign.

The Maturity Scale Gets an Upgrade

In 2016, the nonprofit data maturity scale described a progression that most of the sector had barely begun to climb. Raw data collection at the bottom. Descriptive analytics — reports telling you what happened — as the first meaningful step. Diagnostic analytics using dashboards and alerts to surface why something happened. Predictive analytics using models to forecast what was likely to happen next. And at the top, prescriptive analytics suggesting the optimal decision given everything the data revealed. Fewer than one percent of nonprofits were operating at that level.

Artificial intelligence doesn't replace this scale. It extends it.

Descriptive AI is already table stakes. Reports and dashboards have been automated and accelerated. AI can generate plain-language narratives from data that would previously have required an analyst to write. Useful, but not a competitive advantage today.

Diagnostic AI is where many organizations are starting to see real value. Rather than building a dashboard and hoping someone notices an anomaly, AI detects it automatically and surfaces it the moment it crosses a meaningful threshold. The major donor who hasn't engaged in six months doesn't wait for someone to run a query. The system flags it the day it matters.

Predictive AI is where the sector was focused a decade ago with modeling, and where it is now dramatically more capable. The precision, scale, and speed of AI-powered predictive models exceed what was previously available to all but the largest organizations. But the foundational logic has not changed: you are still predicting the likelihood of a behavior — giving, lapsing, upgrading, making a bequest — based on patterns in historical data. Better models do not rescue poor data. They require better data to produce better predictions.

Generative AI is the genuinely new capability on the scale. The ability to produce original content — communications, summaries, briefings, proposals — at scale and with personalization previously impossible at volume is a legitimate step change. But it introduces a specific risk that Pierce would have named immediately: content deployed without meaningful human review is not personalization. It is the appearance of personalization. Donors can tell the difference.

The organizations that use generative AI well treat it as a drafting and scaling tool, not a substitute for authentic voice and genuine relationship.

Agentic AI — systems that take sequences of actions autonomously in service of a goal — is the emerging frontier. Automating a cultivation sequence, generating a briefing, updating a portfolio record, and scheduling a follow-up visit without human intervention represents real capability.

It also represents real risk without appropriate oversight. Pierce warned against staff-ization — paying people to do everything and losing the human authenticity that makes fundraising work. Agentic AI is the modern equivalent risk: automating everything and losing the human judgment that makes relationships genuine. The organizations navigating this well are doing so with careful experimentation, explicit boundaries, and a clear sense of which decisions require a human in the loop.

The maturity scale is longer now. The distance from the bottom to the top has grown. But the principle anchoring every level is the same one Pierce identified in 1932: everything depends on what you build the foundation on.

Seven Culture Types for the AI Era

Data Driven Nonprofits identified seven distinct culture types among organizations that successfully became more data driven. They weren't stages on a linear progression — an organization didn't have to master one before achieving the next. They were different dominant orientations, each creating its own pathway forward.

The same seven culture types determine AI readiness today. What has changed is the urgency and the specific behaviors each type now demands.

Culture of Champions. Ten years ago, champions were leaders and influential staff who embraced data, modeled its use, and created permission for analytical work. Today the most important champions for AI aren't the most technically sophisticated people in the room — they're the leaders who ask the harder questions Pierce would have asked.

Not *how do we implement AI* but *what problem are we actually trying to solve, and is AI the right tool for it?* Pierce's ideal leader was someone with perspective — someone who could "see the whole field and clearly understand what remains unfinished."

That quality of perspective is exactly what AI championship requires. Champions who lead with honest evaluation rather than uncritical enthusiasm are the ones who set organizations up for sustainable adoption. Champions who lead with enthusiasm alone create expensive decoration.

Culture of Testing. Ten years ago, a testing culture meant A/B testing appeals, piloting new segmentation approaches, and accepting that short-term failure was the price of long-term improvement. Today a culture of testing is more critical for AI than for any previous technology, because AI systems require ongoing evaluation in ways that static reports never did.

Pierce's tested methods principle — number ten on his list of campaign success factors — wasn't about finding one approach and sticking with it forever. It was about learning through disciplined repetition what actually works versus what merely seems to work.

The same discipline applied to AI means treating every deployment as a hypothesis to be tested, not a solution to be celebrated.

Culture of Change. Ten years ago, a culture of change meant tolerance for calculated risk and willingness to disrupt established practices when evidence pointed to better options. Today AI requires something more fundamental. It challenges assumptions about which work requires human judgment and which can be augmented by technology. Pierce navigated this tension throughout his career — he brought systematic methods to fundraising precisely because he believed better systems freed people to do more important human work.

That's the right frame for AI adoption too. The hardest change isn't deploying AI. It's reimagining workflows in ways that use AI to do more of what matters, not less of what works.

Culture of Sharing. Ten years ago, data silos were the enemy of data-driven decision making. Organizations with a sharing culture had dismantled walls between departments, established shared infrastructure, and created norms of information transparency.

Today AI systems require shared data to function at all. A major gift model that only sees gift officer visit data will miss the signals visible in annual fund giving patterns. A donor retention model that only sees the fundraising database will miss engagement signals captured by the program team.

Organizations that have broken down silos hold a structural advantage. Those that haven't will find their AI is only as good as their most isolated data source.

Culture of Growth. Ten years ago, growth culture meant benchmarking rigorously, tracking metrics consistently, and setting goals that were genuinely stretching.

Today AI creates extraordinary growth opportunities — but only for organizations that define growth the way Pierce would have: in terms of mission outcomes, not tool adoption. Pierce warned about organizations that became complacent when resources felt comfortable — where vision disappeared among satisfied trustees and dynamic programs deteriorated into self-satisfied routine. The AI equivalent is measuring AI adoption itself: tools deployed, features activated, content volumes generated.

A genuine growth culture holds AI to the same standard it holds every other investment: does it improve donor retention, increase major gift conversion, grow planned giving commitments? Those are the numbers that tell you whether AI is advancing the mission or merely decorating it.

Culture of Agile. Ten years ago, agile culture meant moving faster and empowering people to make decisions without top-down approval at every step.

Today the pace of AI development demands agility most nonprofits haven't developed yet. What was state-of-the-art twelve months ago is already being superseded. An agile culture engages with AI the way Pierce approached tested methods — with ongoing evaluation, willingness to adapt, and the discipline to abandon approaches that aren't delivering regardless of the initial investment.

Rigidity in either direction — wholesale resistance or premature commitment — is equally dangerous.

Culture of Data. Ten years ago, a culture of data meant data was a genuine and respected input into strategy and operations. Today this is the foundational culture type, and it carries more weight than ever.

Pierce built his entire methodology on one foundational conviction: know your field before you act in it. For AI, knowing your field begins with knowing your data.

Organizations that haven't built a genuine culture of data — where data health is maintained rigorously, where analytics inform decisions at every level, where the No Garbage Rule is practiced rather than posted — aren't ready for heavy AI investment.

They should be investing in data first. AI built on a poor data culture will not solve the underlying problems. It will amplify them, with speed and confidence that make the errors harder to detect and harder to walk back.

The ABCs of AI Culture Adoption

Building a culture of AI adoption follows the same logic as building a culture of data use. The same three principles that guided data culture adoption a decade ago apply here with equal force.

Acknowledge the Current Culture. Pierce was direct about this: you cannot build a successful campaign — or anything else — on a foundation you haven't honestly assessed. Start with an honest evaluation of where your organization actually is.

Are you genuinely data driven today? If the answer is mostly yes — your data is reasonably healthy, analytics inform key decisions, your team is comfortable working with evidence — then your foundation for AI is solid.

Build from that strength. If the honest answer is that your organization is still working toward data-driven practices, name that clearly. Not as an indictment, but as a starting point.

You cannot use AI to leapfrog data maturity. The organizations that have tried have mostly created expensive and confusing experiences for staff and donors alike.

Baby Step Behaviors. Pierce understood that great administrators make themselves dispensable by building forces that can carry on. "Visionary leaders building real foundations," he wrote, "seek to make themselves dispensable by creating forces that can carry on." That principle applies directly to AI culture building.

Don't attempt to transform the entire organization's relationship with AI at once. Find the early adopters and give them specific, bounded applications to test. Choose people who are networked across the organization, so when they share what they're learning, it travels. Include at least one skeptic, and support them through a genuine trial.

When the skeptic's experience is positive, the cultural ripple effect is far larger than when only the enthusiasts succeed.

Target behaviors that are observable, repeatable, and measurable — a gift officer using an AI-assisted prospect briefing before every major visit and reporting better conversations is creating the kind of visible behavior change that builds culture from the ground up.

Culture Aligns with Strategy. Pierce's eleven success factors weren't a checklist to work through arbitrarily. They were a framework for aligning organizational capacity to a specific, defined objective.

The most important alignment question for AI is the same one Pierce asked before any campaign: What are we actually trying to accomplish, and do we have the organizational strength to accomplish it? If the goal is major gift growth, AI applications that improve prospect identification and visit preparation are directly strategic.

If the goal is donor retention, AI-powered lapse prediction and personalized re-engagement are the priority. If the goal is planned giving development, using AI to identify legacy indicators across a large donor file is the obvious first application. Organizations that adopt AI without connecting it to specific strategic priorities end up with tools that generate activity without producing impact. The strategic question comes first. The AI application follows.

Two Organizations, One Fork in the Road

Chapter One described the Nonprofit AI Divide — organizations moving fast versus organizations paralyzed. Pierce described the same divide a century earlier, though he framed it differently. Some organizations, he observed, had leaders with burning conviction, clear perspective, and the organizational discipline to translate both into results. Others had resources, intentions, and even good programs — but lacked the cultural foundation to execute at their potential. The same distinction holds today. There are two types of organizations building genuine AI capability.

The first group spent the decade between 2016 and 2026 doing the hard work of data culture. They invested in data health. They built analytical capabilities. They developed internal champions. They established norms where evidence informed decisions at every level. For them, AI isn't a cultural transformation — it's a technological evolution of practices already in place. Their data is healthy enough to learn from.

Their staff is analytically literate enough to evaluate AI outputs critically. Their leadership has enough experience with data-informed decisions to ask the right questions of AI-generated recommendations.

These organizations are not the majority. But they exist across every part of the sector, and they demonstrate that a decade of disciplined investment in data culture creates genuine structural advantage when the next generation of tools arrives.

The second group is the larger cohort. They may have CRM systems and analytics platforms. They may have run predictive models at some point. But they never truly built the culture.

Decisions still run primarily on instinct and relationship. Data health gets addressed reactively. Analytics get produced for leadership on request rather than embedded in daily workflows.

For these organizations, the path to becoming AI-driven runs through exactly the same territory the path to becoming data-driven did. There are no shortcuts. But the roadmap is clearer now than it was a decade ago.

We know what works. We know what first steps look like. We know that organizations of every size, cause, and budget level have successfully made the journey.

The first step is the same one Pierce would have prescribed. Acknowledge where you are. Accept it without flinching. Know your field as it actually is, not as you wish it were. And then decide to take one step forward.

What First Steps Actually Look Like

The first steps to building an AI-driven culture are measured, specific, and focused on building habits rather than deploying technology — exactly the kind of "unhurried period of preparation" that Pierce identified as a critical success factor for any serious undertaking.

Start with an honest data health assessment. Before evaluating any AI platform or capability, get a clear picture of where your data actually stands. What percentage of your records have valid addresses? Valid email addresses? What does your duplication rate look like? When was your wealth screening data last refreshed? These aren't interesting questions — they're essential ones.

The answers tell you which AI capabilities are accessible right now and what investment you need to make before others become viable. Pierce called this knowing your field. It's still the right place to start.

Identify one specific problem AI could address. Not a vague ambition like *improve donor communication* but a specific, measurable problem: we lose twelve percent of our $500-to-$999 donors annually and can't identify them in time to intervene. Or: our gift officers spend six hours preparing for each major donor visit and we need to cut that substantially without reducing quality.

Pierce built his campaigns around specific, defined objectives — a reasonable and clearly articulated goal was number three on his success factors list.

The same discipline applies here. Specific problems lead to specific applications. Specific applications can be evaluated against clear success criteria. Vague ambitions lead to pilot programs that produce unclear results and get quietly discontinued.

Build in the human review loop from the beginning. Whatever AI application you deploy, design human checkpoints into the workflow deliberately — not as an afterthought or a concession to staff anxiety, but because the human review loop is where AI outputs get corrected, refined, and grounded in relational context no algorithm can access.

Pierce warned against staff-ization specifically because it removed the human authenticity that made fundraising effective. The human review loop in an AI workflow serves the same protective function.

A gift officer reviewing an AI-generated prospect briefing before a visit isn't auditing the technology. They're bringing the judgment, context, and relationship knowledge that converts information into action.

Measure what matters, not what is easy. AI tools generate a lot of activity metrics — messages sent, scores calculated, prospects flagged, reports generated. These are easy to track and largely meaningless in isolation. Pierce's eleventh success factor was competent direction — and competent direction means knowing whether the campaign is actually moving toward its goal, not just whether activity is occurring. The metrics that matter are the same ones that mattered before AI arrived: donor retention rates, dollars raised, major gift conversion, planned gift commitments.

If AI is working, these numbers improve. If they're not improving, no volume of activity metrics changes that honest conclusion.

Culture Is the Strategy

Chapter One made the case for urgency. The Nonprofit AI Divide is real. The gap between organizations that adapt and those that don't is growing exponentially. The window for building foundational culture before AI becomes a competitive necessity is narrowing.

This chapter is the answer to that urgency. Not a faster deployment schedule. Not a bigger technology budget.

Culture.

Pierce identified eleven factors that determine campaign success before a campaign even begins. Not one of them is about the sophistication of the tools available. Competent leadership. A compelling cause. Tested methods. An informed constituency. These are culture factors. They describe what an organization is, not what it uses.

McLuhan wrote that we shape our tools and thereafter our tools shape us. The nonprofit organizations that will look back in a decade and recognize themselves as genuinely AI-driven will be the ones that did the shaping.

They asked what problem they were trying to solve before they asked which tool might solve it. They invested in data health before AI infrastructure.

They built cultures of testing, sharing, and growth that made every new capability sustainable. They kept the human being at the center of every donor relationship — and used AI to make those relationships more informed, more timely, and more personal without making them less real.

Pierce built the art of fundraising without donor databases, without analytics platforms, without email, without websites, without artificial intelligence.

What he had was systematic thinking, a culture of accountability, an unshakeable belief that the cause was worth the effort, and a burning conviction that the work had to be done.

That's still the foundation. Everything else — including AI — is built on top of it.

04 ANXIETY

"The oldest and strongest emotion of mankind is fear, and the oldest and strongest kind of fear is fear of the unknown."
— H.P. Lovecraft

Knowns and Unknowns

Walk into any nonprofit conference session about AI and you will witness a familiar human drama playing out in real time. In the front rows, the enthusiasts lean forward. They are already using AI tools. They have stories about time saved, content generated, prospects identified. They want to talk about what comes next.

In the back rows, or just outside the door entirely, is everyone else. They are not indifferent. They are anxious. Some are anxious that their organization is falling behind. Some are anxious that AI will eliminate jobs — theirs, or their colleagues'.

Some are anxious about doing something wrong. Saying something wrong. Making a decision with a tool they don't fully understand and can't fully explain to their board.

And some are anxious about something harder to articulate — a vague but persistent feeling that something important is at stake in how the nonprofit sector answers the AI question.

That the values that define this work — human dignity, authentic relationship, service to mission — could be quietly compromised in the rush to keep up.

That anxiety is not irrational. It is not ignorance. In many cases it reflects exactly the kind of careful thinking that nonprofit leaders are supposed to bring to decisions that affect the people they serve.

This chapter takes that anxiety seriously. Not to dismiss it. Not to explain it away with reassurances about how AI is just a tool. The concerns are real, some of the risks are real, and the leaders who are asking hard questions deserve honest answers — not a sales pitch dressed up as strategy.

But there is a difference between anxiety that leads to useful caution and anxiety that leads to paralysis. The goal here is to name what is actually worth worrying about, separate it from what is noise, and give nonprofit leaders the framework to make decisions with open eyes.

We Have Been Here Before

The anxiety surrounding AI is genuine. It is also not new. Every significant technology introduction into the nonprofit sector has generated the same arc: breathless enthusiasm from early adopters, genuine concern from thoughtful skeptics, organizational paralysis in the middle, and eventually — sometimes years later — a kind of embarrassed consensus that the technology was neither as revolutionary nor as dangerous as either camp originally claimed.

Roger Craver, one of the pioneers of direct mail fundraising, tells the story with characteristic directness. In the early 1970s, the prevailing wisdom in the sector was that direct mail was fundamentally incompatible with serious philanthropy. The saying among major gift professionals was blunt: "You can't get milk from a cow by sending it a letter."

The anxiety was real. Direct mail felt impersonal. It felt transactional. It felt like it would cheapen the donor relationship.

Then, in the summer of 1970, Common Cause launched. The organization went from zero to 500,000 dues-paying members in a year through direct mail. The herd, as Craver puts it, followed. Within a decade, organizations that had resisted direct mail were entirely dependent on it.

The same arc played out with database systems in the 1980s. With online giving in the late 1990s — remember when the prevailing wisdom was that donors would never trust putting a credit card number into a website? Online giving in the United States now exceeds $60 billion annually. With social media in the 2000s. With mobile fundraising in the 2010s. With data analytics in the years surrounding the publication of *Data Driven Nonprofits* in 2016.

Each wave generated anxiety. Each wave produced a cohort of early adopters who moved too fast without enough judgment, a cohort of thoughtful practitioners who navigated the transition carefully, and a cohort of resisters who eventually found themselves using the technology anyway, years behind where they could have been.

The nonprofit sector is not unique in this. Gartner's Hype Cycle — the research framework that maps how technologies move from inflated expectations through disillusionment to productive use — describes a pattern that has repeated reliably across nearly every significant technology introduction of the past forty years.

The peak of inflated expectations is where AI sits right now in many corners of the sector. Every problem looks solvable. Every vendor has a miracle. Every conference session promises transformation.

The trough of disillusionment comes next. Some organizations will have moved fast on bad foundations, produced poor results, and concluded that AI doesn't work — when the actual lesson is that AI doesn't work on broken data and without strategic intent.

The organizations that survive the trough with their enthusiasm intact will be the ones that entered with honest expectations. The organizations that reach what Gartner calls the plateau of productivity — where AI becomes genuinely integrated into how work gets done — will be the ones that approached the technology the way Pierce approached tested methods: with ongoing evaluation, willingness to adapt, and the discipline to abandon approaches that aren't delivering.

The anxiety is familiar. The arc is predictable. What is not inevitable is where your organization ends up.

Crossing the Chasm in the Nonprofit Sector

Geoffrey Moore's *Crossing the Chasm* introduced a concept that has outlasted most business books of its era: the idea that there is a dangerous gap between early technology adopters and the mainstream market, and that most technologies fail in that gap rather than crossing it.

Early adopters are a specific personality type. They are comfortable with incomplete information. They make decisions based on vision rather than proven results. They tolerate failed experiments as the cost of learning. They are often wrong in the details and right about the direction.

The early majority — the mainstream — is different. They want to see references. They want to know that an organization like theirs, facing problems like theirs, with resources like theirs, used this technology and it worked. They are not risk-averse in the pejorative sense. They are responsible stewards of organizational resources making rational decisions under uncertainty.

For AI in the nonprofit sector, the chasm is visible and specific. Early adopters are already there. They are the large research universities running AI-powered prospect identification models across millions of alumni records. The national health systems using predictive analytics to identify grateful patients before they are ever approached. The digital-native nonprofits that have built AI into their operations the way a tech company would.

The mainstream — the tens of thousands of community foundations, regional nonprofits, social service organizations, advocacy groups, and educational institutions that make up the backbone of the sector — is watching from the other side of the chasm, waiting for something that looks more like proof than promise.

What crosses the chasm in the nonprofit sector is not hype. It is not conference presentations from early adopters who have more resources and technical capacity than most organizations will ever have.

What crosses the chasm is when an organization of recognizable size, with recognizable resource constraints, in a recognizable sub-sector, solves a recognizable problem with AI and talks about it plainly.

The organizations on the far side of the chasm right now are not waiting because they are timid. They are waiting because the bar for sufficient evidence hasn't been met yet — and in the context of serving constituencies who depend on the organization's continued effectiveness, that is a defensible position.

The acceleration happens when that evidence accumulates. When the mid-size human services organization in the Midwest shares what it cost, what it took, what it produced, and what it would do differently. When the small community foundation talks about how it used AI tools that were accessible and affordable and specific about what they did and didn't deliver.

When the stories that circulate at sector conferences start to sound like the organizations in the room rather than the outliers.

That accumulation is already beginning. The chasm is not uncrossable. But it requires honesty from the organizations on the early side about what is actually replicable — not just what is impressive.

The Concerns That Are Worth Taking Seriously
Not all AI anxiety is the same. Some of it is fear of change dressed up as ethical reasoning. Some of it is legitimate concern about real risks that deserve real attention.

The difference matters, because dismissing the latter along with the former is one of the ways organizations end up making decisions they later regret.

Here are the concerns that deserve honest engagement.

Bias in AI systems is real.
AI models learn from historical data. Historical data in the nonprofit sector reflects historical decisions — about who got contacted, who got cultivated, who got asked, who gave, and who didn't. If those decisions were biased — by race, by geography, by socioeconomic status, by any of the ways institutional patterns get embedded in data — then an AI model trained on that data will reproduce and potentially amplify those biases.

This is not a theoretical concern. It is a documented phenomenon in every sector that has deployed AI systems at scale. Hiring algorithms trained on historical hiring data replicated historical hiring preferences. Credit scoring models trained on historical lending data reproduced historical lending disparities. There is no reason to believe nonprofit fundraising algorithms are immune.

The appropriate response is not to avoid AI. It is to examine what the model is learning from, audit the outputs for patterns that reflect bias rather than merit, and build in the human review that catches what the algorithm misses.

A major gift identification model that surfaces a pattern of only flagging prospects from wealthy zip codes deserves scrutiny — not because the model failed, but because it may be surfacing a historical pattern that deserves to be disrupted rather than reinforced.

Privacy concerns are legitimate and specific. AI-powered fundraising tools aggregate data from multiple sources: giving history, engagement records, wealth screenings, public records, social media signals, demographic data. The picture that emerges can be remarkably detailed.

Donors who provided their name and address for a newsletter did not necessarily consent to becoming the subject of a detailed predictive profile.

This is not an argument against data-informed fundraising — it is an argument for being thoughtful about what data is used, how it is used, and what disclosure obligations exist. Most donors understand and accept that organizations they support will use reasonable information to communicate with them.

Most donors do not know the extent of external data being appended to their records. There is a transparency gap worth closing.

The reasonable standard: use data in ways that would make donors feel known and valued if they were aware of it. Donor-centric data use is data use in service of the relationship. Data use that would feel invasive or surveilling if donors understood it fails that standard regardless of whether it is technically permissible.

The job displacement question is real, even if the answer isn't clear. The nonprofit sector employs millions of people, many in roles that involve data processing, administrative coordination, content creation, and research — functions where AI is already demonstrably effective.

The anxiety that AI will eliminate development associate positions, grant writer roles, communications coordinator functions, and administrative staff roles is not unfounded. The honest answer is that some roles will change significantly.

Some functions currently performed by people will be automated. This is already happening, not at some distant future date. The organization that pretends otherwise is not reassuring its staff — it is setting them up for a harder conversation later.

The more useful framing is that AI will change what skills are valuable rather than simply eliminating roles. The development professional who can evaluate and refine AI-generated content is more valuable than one who produces it from scratch.

The analyst who can interrogate a model's outputs and identify where they go wrong is more valuable than one who pulls reports. Organizations that invest in training staff to work alongside AI tools are better positioned than those that either eliminate jobs abruptly or avoid the question entirely.

Donor trust is a genuine consideration. If donors discover that the deeply personal letter they received was entirely AI-generated, that the thank-you call was triggered by an algorithm, or that their giving history was fed into a system they were never told about, some of them will feel manipulated rather than cultivated. Donor trust, once damaged, is expensive to rebuild.

The standard here is authenticity, not prohibition. An AI-drafted communication reviewed and personalized by a real human who knows the donor is not deceptive — it is a tool that enhanced the human's capacity to communicate well.

A fully automated system that removes the human from the relationship while creating the appearance of human attention is a different thing entirely.

The Concerns That Are Mostly Noise

Honesty requires distinguishing legitimate concern from anxiety that generates heat without much light.

The fear that AI will "replace" major gift fundraising is not well-founded. Major gifts have always required, and will continue to require, personal relationships developed over time. The case that a donor who has been cultivated for years makes when writing a seven-figure check involves conviction, trust, and human connection that no algorithm produces.

AI changes the information available to the people doing this work. It does not change the fundamental dynamic of how transformational gifts happen.

The fear that AI will "know too much" and somehow control donor decisions reflects a misunderstanding of how the tools actually work. A predictive model produces probabilities, not certainties.

It suggests priorities; it does not determine outcomes. The gift officer who reviews an AI-generated prospect list is exercising judgment, not executing instructions. The human is still in the room. The human is still making the call.

The fear that AI will permanently dehumanize the nonprofit sector assumes that organizations have no agency in how they deploy tools.

Every AI application is a choice. The organization that decides to use AI to give gift officers better information before donor visits, and the organization that decides to use AI to eliminate the gift officer visit entirely, have made very different choices. The technology is the same. The values embedded in its deployment are not.

Responsible AI: What It Actually Requires
The phrase "responsible AI" has become something of a buzzword — invoked often enough that it risks losing meaning. For nonprofit leaders trying to build actual governance frameworks rather than aspirational language, the concept needs to be operationalized.

Responsible AI in the nonprofit sector requires five things.

Intentional purpose. The responsible AI question begins before any tool is deployed: what problem are we solving, and is AI the right tool for it?

Organizations that adopt AI because of competitive pressure, vendor enthusiasm, or board curiosity — rather than because a specific operational problem warrants the solution — are not in a position to be responsible stewards of the tool. Purpose comes first.

Data governance that precedes deployment. Responsible AI requires knowing what data the system is learning from, where that data came from, whether it has been maintained with sufficient quality, and whether its use is consistent with the organization's values and obligations to constituents.

This is not a technical question delegated to IT. It is a leadership question. The chief development officer who cannot answer basic questions about what data feeds the AI systems making recommendations about her donors has a governance problem.

Transparency with stakeholders. Responsible AI doesn't require disclosing every technical detail of every system to every donor. It does require not actively deceiving people about how the organization uses information and technology.

The privacy policy that was last updated in 2015 and says nothing about AI is not adequate. The staff handbook that describes data processes as they existed before AI integration is not adequate. Reasonable transparency is not difficult. It requires being honest about what you're doing and why.

Human oversight as a design principle. The most important safeguard in any AI system is the human being who reviews its outputs before they drive decisions or actions. This is not an afterthought. It should be designed into every workflow.

The gift officer who reviews AI-generated prospect prioritizations before acting on them. The communications director who reads AI-drafted donor communications before they go out. The development director who audits model outputs quarterly for patterns that suggest bias or error. Human oversight is not inefficiency. It is accountability.

Ongoing evaluation against mission. Responsible AI is not a one-time governance decision. It is a continuous practice of asking whether the tools are delivering what they promised, whether they are creating unintended consequences, and whether the organization's use of them remains aligned with its values. Pierce's standard applies here as everywhere: is the effort actually moving toward the goal?

What an AI Policy Actually Needs to Cover

Many nonprofit organizations are now being asked — by boards, by funders, by staff — to produce an AI policy. Most of the policies being drafted are either so broad they say nothing useful, or so restrictive they prohibit things that are already happening and probably should happen.

A useful AI policy for a nonprofit organization needs to address several specific areas.

Approved use cases. What AI applications has the organization decided to adopt, and why? This is not a comprehensive list of every tool staff may ever touch — it is a documented rationale for the organizational AI investments that have been made. Which problems are being addressed. What success looks like. What the review process was before adoption.

Staff use guidelines. As AI tools become embedded in commercial software — CRM systems, email platforms, document editors, communication tools — it becomes increasingly difficult to prohibit staff AI use entirely, and attempting to do so creates a different problem: staff who use AI tools anyway, without disclosure, without training, and without quality review.

A more useful policy defines what responsible staff use looks like. When human review is required. What cannot be delegated to AI tools. What disclosure is expected when AI-generated content is used.

Constituent data governance. What data about donors, volunteers, clients, and other constituents may be used in AI systems? What data may not? What third-party AI tools are permitted to process organizational data, and under what terms?

This section requires actual engagement with the organization's data, its existing privacy commitments, and the terms of service of the tools being considered. It is not glamorous work. It is essential work.

Vendor assessment criteria. How does the organization evaluate AI vendors and tools before adoption? What questions about data practices, model transparency, bias testing, and security are required? What is the process for ongoing monitoring after adoption?

Organizations that adopt AI tools without this framework are outsourcing governance decisions to vendors whose interests are not identical to the organization's.

Prohibited uses. What will the organization not do with AI, regardless of capability? This list should be short and specific — a genuine ethical line rather than a collection of vague concerns.

Examples that belong on this list: using AI to generate donor communications that are represented as personally written by leadership when they are not reviewed by that leadership; using AI outputs without any human review in decisions that materially affect individuals; using AI tools that process sensitive constituent data without appropriate data protection agreements.

The Question Nonprofit Leaders Need to Answer

Behind all of the specific concerns, the hype cycle analysis, the governance frameworks, and the responsible AI language, there is a simpler and harder question that every nonprofit leader eventually has to answer.

The question is not: Is AI safe?

The question is: What is this organization's obligation to the people it serves?

If the answer is to serve as many people as effectively as possible with the resources available, then the AI question resolves fairly directly. Tools that expand capacity, improve decision-making, and make limited resources go further in service of mission are worth understanding and — where appropriate — adopting.

The anxiety that prevents organizations from getting there is understandable. The sector was built on human relationship. It attracted professionals who believe in the primacy of personal connection.

It serves constituencies who are often vulnerable in ways that demand extra care. The worry that technology will cheapen this work, reduce it to algorithms, make it feel like commerce rather than care — that worry comes from a good place.

But the organizations serving the most people, retaining the most donors, identifying the most planned giving prospects, getting the most out of every dollar invested in development operations — those organizations are not choosing between human values and AI capability.

They are using AI to extend the reach of their human values. To find more people who care. To have better conversations with them. To steward the relationships that result with more attention and more care than they could provide without the help.

Pierce built a fundraising methodology on the conviction that the cause was worth fighting for and that the tools available — however limited — should be used in full service of that fight.

He would have no patience for the anxiety that refuses to act. He would have equal impatience with the enthusiasm that acts without judgment.

The goal is neither. It is clear-eyed, mission-grounded, values-anchored movement forward.

The anxiety is a signal worth listening to. It is not a destination.

The Things That Don't Change

Every technology transition in the sector has ultimately confirmed the same set of durable truths that Ward and Pierce discovered at the beginning of the twentieth century.

People give to causes they care about, presented by people they trust. Major gifts require personal relationships developed over time. Authentic conviction is more persuasive than eloquent technique. Stewardship determines whether donors come back. The cause is always larger than the tool.

AI will not change these things. It will not reduce the importance of the gift officer who walks into a living room and makes an ask with genuine belief in what the organization is doing. It will not make the thank-you call less meaningful. It will not replace the board member who calls her friend because she believes in the work.

What it can do — what it is doing, in the organizations that are getting this right — is give every person doing this human work better information, more time, and greater capacity to focus on the parts of the job that only humans can do.

The anxiety is about whether that trade-off is real. Whether the efficiency gains actually free up human capacity, or whether they just become justification for reducing it. Whether the personalization is genuine or performed. Whether the tools serve the mission or gradually redirect it.

Those are the right questions. They are worth asking loudly, repeatedly, and without embarrassment.

05 FLAWS

"The most dangerous phrase in the language is, 'We've always done it this way.'" — Rear Admiral Grace Hopper

Honest Questions

There is a conversation that happens in development offices across the country with remarkable consistency. A staff member — or sometimes a board member — hears about predictive modeling and asks, reasonably enough: "Can't we just pull a report and sort by giving history? Can't we build a pivot table for that? Can't we use RFM?"

These are honest questions. They deserve honest answers. No, no, and not anymore. And understanding precisely why you can't is one of the most important things a nonprofit leader can grasp about how modern fundraising actually works.

The people asking these questions are not lazy or uninformed. They are working hard, operating with limited resources, and trying to make good decisions for their organizations. Their intentions are exactly right. The tools they are reaching for are exactly wrong — not because the tools are useless, but because they are being asked to do something they were never designed to do.

Grace Hopper spent decades watching smart, well-meaning people defend familiar methods simply because those methods were familiar. She understood that institutional comfort with a tool is not the same thing as the tool being right for the job.

The danger she named isn't stubbornness — it's something more subtle and more common. The reasonable assumption that because something has worked well enough for long enough, it must still be working well enough now.

In nonprofit fundraising, that assumption has a name. It's called RFM. And it has reached its limits.

Start With What You're Actually Trying to Do

Let's define the problem clearly, because clarity matters here. You have a list of prospects. Some of them are going to make a gift. Some of them aren't. Your job — at its most fundamental — is to figure out which is which, and direct your limited time, staff, and budget toward the people most likely to give.

At fifty prospects, a skilled gift officer can hold most of that in their head. They know who came to the last event, who responded to the last appeal, who seems warm when called.

Human pattern recognition works reasonably well at that scale.

Now scale that up to 5,000 prospects. Then 50,000. Then 500,000.

At some point — and that point arrives faster than most people expect — the tools break before the people do. Not because the staff aren't talented. Because the problem is mathematically beyond what any report, pivot table, or three-variable segmentation framework can solve.

What a Report Can Tell You (And What It Can't)

A report is a window into the past. It answers one question: *What happened?*

Pull a report on your donors from the last three years and you can see who gave, when they gave, how much they gave, and through which channel. Useful information. The starting point for most development shop analysis, and there's nothing wrong with it as far as it goes.

The problem is where it stops.

A report can tell you that 847 people gave between $500 and $2,500 last year. It cannot tell you which of the 4,153 people who didn't give at that level have the capacity and inclination to do so this year.

It describes your known donors. It tells you nothing useful about your unknown ones — the people in your database who should be in your pipeline but aren't yet.

Descriptive analytics — which is what reports give you — describes a population you already know. It cannot predict behavior in a population you don't. The donors you most need to find are outside its field of view by definition.

What a Pivot Table Can Tell You (And What It Can't)
A pivot table is a more sophisticated version of the same problem.

You can slice your data. Group donors by giving level, geography, event attendance, consecutive years of giving. Create segments and cross-reference them.

A skilled analyst can build a pivot table that surfaces genuine patterns — donors who attend events give at higher rates, five-plus year consecutive givers show strong retention profiles, online donors respond better to email.

That's real and useful. But here's what a pivot table cannot do: weigh dozens of variables simultaneously, assign relative importance to each one, account for interactions between them, and produce a ranked probability score for every individual record in a database of 50,000.

A pivot table works with the variables you choose to look at, one combination at a time. A predictive model works with every variable at once — including interactions between variables that no human analyst would think to examine. Those are not the same instrument. They are not even close.

What RFM Can Tell You (And What It Can't)
Before we talk about what predictive models can do, we need an honest conversation about Recency, Frequency, and Monetary value — the framework that has dominated nonprofit fundraising segmentation for decades, and the one that most development professionals reach for first.
RFM deserves credit.

It became the industry standard for a reason, and Chuck Longfield — one of the most important figures in nonprofit data science over the last four decades — explains that reason precisely: "RFM dominates because I can visualize what I'm doing. The variables are directional: $10 is better than $5, $25 is better than $10. Two gifts are better than one gift. If I gave a year ago, that's better than giving my last gift two years ago. I can chop them up into segments and know this segment is better than that segment — and if I have to cut some segment out because I don't have the money to send all that mail, I know where to cut."

That's not a small thing. Recency, Frequency, and Monetary value have two characteristics that make them uniquely manageable for human analysis. They have *continuity* — the values don't jump around randomly. And they have *directionality* — more is always better.

Those relationships are stable, intuitive, and easy to hold in mind simultaneously. For a world where mail costs money and you need to make triage decisions about which segments get solicited, RFM is a logical and practical framework.

It was borrowed from direct marketing, applied to nonprofit fundraising, and it worked well enough for long enough that an entire professional generation built their workflows around it.

But here is the problem Longfield identified — and it deserves to be read slowly: "That's RFM, but people have immense amounts of data in their systems and they have had it for decades. They know whether the person volunteers or gives through multiple channels. They know whether the person has opened an email or called the call center. Yet, the tool through which all this data must be funneled to create value can't handle it."

That is the Nonprofit Data Paradox in its clearest expression. The more data you have, the less valuable it becomes — not because the data isn't good, but because the analytical tool most organizations rely on can't use it. Something that should increase exponentially in value actually decreases at a very precipitous rate.

The Four Structural Flaws in RFM

RFM isn't wrong. It's incomplete. And at scale, incomplete becomes a meaningful liability.

Flaw 1: It ignores everything that isn't giving history. RFM, by definition, only looks at what someone has given to you — when, how often, and how much. Your database almost certainly contains far more than that. Volunteer history. Event attendance. Email engagement. Call center contacts. Website behavior. Whether someone serves on the board of another nonprofit. Whether they've made gifts to peer organizations. Whether their wealth profile has changed since the last screening.

None of that makes it into an RFM calculation. The framework was designed for simpler data environments than the ones most organizations now operate. Using RFM in a modern CRM is like navigating a city with a map drawn thirty years ago. The roads you know are still there. Everything that's been built since is invisible to you.

Flaw 2: The tool can only manage a handful of variables simultaneously.
Longfield is direct about this ceiling: "Nobody's going to be able to type the query or write the memo that says to the database administrator, 'I want to take the people who are male, who came to the website twice, who called the call center, who gave between $50 and $100, and on and on and on. I want to send them this letter and I want to ask them for this amount of money through this channel.'"

Three variables are manageable. Five become difficult. Ten become impossible. Twenty becomes fantasy. RFM persists not because it's the best available segmentation but because it's the most complex segmentation that humans can reliably execute without a model. That's a cognitive limitation built into the tool's design — not an analytical achievement.

Flaw 3: It treats giving history as a proxy for giving potential — and those are not the same thing.
A donor who has given $150 per year for fifteen years has a strong RFM profile — high recency, high frequency, moderate monetary value.

But RFM cannot tell you whether that person has the capacity to give $15,000, or whether they're already giving $50,000 to three other organizations and $150 is genuinely a stretch. It scores what people have done. It has no mechanism for assessing what they could do.

This is where external data becomes decisive, and where RFM's internally-focused view becomes a genuine strategic problem. The average American who gives to charity supports between four and eight organizations per year. Your RFM score is based entirely on what they gave *you*. It's a partial view being treated as a complete picture.

Longfield describes the missed opportunity precisely: "If I'm trying to do the ask string and I want someone to give more money, it would be nice to know how wealthy they are, but I also need to know how that was correlated. If I simply wanted to set the ask string based on your last gift of $100 and say $100, $200, $300 — if you're wealthy, maybe I want to use $110, $220, and $330. But I can't do that with segmentation."

Flaw 4: It misses the donors you most need to find.
This is the most consequential flaw of all, and it's structural. RFM can only evaluate and rank the people already behaving in ways that show up in your giving records. It cannot surface the prospect who has never given to you but looks exactly like your best major donors in every way that matters — wealth, philanthropic history, affinity to your mission, connections to your community.

Longfield gives the university example: were they on financial aid and feel indebted to the institution? Did they play varsity sports? Is their family a legacy? Every one of those things would adjust an ask string — make you a little more aggressive, or mail someone one more time than people who didn't share those characteristics.

These correlations are knowable and predictable. They are not random. But they appear in institutional records that have sat in the database for decades without ever being connected to fundraising strategy, because RFM has no way to reach them. The donors you don't know you should know are invisible to RFM. They're hiding in your data right now.

The Scale Problem, Made Concrete

Here is where the math becomes impossible to ignore. Imagine your database has 50,000 records. You want to know which of those people have a meaningful likelihood of making a gift in the next twelve months — not who gave before, but who *will* give, including people who haven't given recently or ever.

A predictive model evaluating those 50,000 records might consider thirty or forty variables for each person. Giving history, yes — but also the trend line of giving over time, event attendance, email engagement, wealth screening data, age and demographic indicators, external charitable giving data, real estate holdings, business affiliations, nonprofit board service, and more.

Thirty variables across 50,000 records is 1.5 million data points before the model starts looking at how those variables interact with each other.

No report surfaces that. No pivot table processes it. No RFM calculation captures it. A good model does it in minutes. It produces a score — typically on a scale of 0 to 1,000 — for every single record.

The prospect with a score of 923 looks very different from the prospect with a score of 312, even if their most recent gift amounts were identical and their RFM profiles were nearly indistinguishable.

The model is incorporating everything the standard toolkit missed.

What the Model Actually Does

A predictive model doesn't predict the future. That's worth stating plainly, because the word *predictive* creates unrealistic expectations in some audiences and irrational skepticism in others.

What a model does is this: it looks at the characteristics of the people who *did* make gifts — their behavioral patterns, financial signals, engagement history, and demographic attributes — and then identifies which of your non-giving or under-giving records share those same patterns. It is finding look-alikes. It is doing at massive scale exactly what a skilled gift officer does intuitively at small scale when they say, "This person reminds me of someone who became a major donor."

As Longfield describes it, a model can determine "what the correlation is of each of these variables to the thing that they're trying to predict or generate." Those correlations aren't random. They're not coincidence. They're knowable and repeatable patterns embedded in your data — patterns that have been there for years but invisible to every report, pivot table, and RFM analysis you've ever run.

The output is a ranked list. The prospect with a score of 950 doesn't *definitely* give. The prospect with a score of 200 doesn't *definitely* not give. What the scores tell you is where to concentrate your finite capacity — where to deploy your staff, where to direct your appeals, where to invest in cultivation.

At 5,000 records, the top-scored 500 represent your highest-probability segment. At 50,000, the top 5,000. At 500,000, the top 50,000. In every case, you've moved from an undifferentiated mass of prospects to a ranked, actionable priority list — one that would have been impossible to produce through manual analysis, and one that RFM alone would have gotten materially wrong.

The Return Is Measurable
This isn't theoretical.

In *Data Driven Nonprofits*, it was noted that the University of South Dakota Foundation tracked every major gift closed against their predictive model scores. The $15,000 invested in one modeling service produced $58.8 million in closed gifts.

That number exists because someone measured the connection between model output and fundraising results — conversion rates, dollars raised from model-predicted prospects versus non-predicted. A 59% discovery visit conversion rate from model-identified prospects doesn't happen because the algorithm is magic.

It happens because the algorithm is surfacing the right signals, and a gift officer is converting those signals into relationships. The model doesn't raise the money. The model tells you where to go. The gift officer raises the money.

That distinction matters enormously. Organizations that deploy predictive models well don't eliminate their development staff — they make their development staff dramatically more productive.

The same number of people, working the same number of hours, generating materially better results because they're directing their effort toward the right prospects rather than working a framework designed for a simpler data environment.

The One Thing That Breaks the Model

There is a catch, and it deserves its own paragraph. Everything described above assumes the underlying data is reasonably clean.

A predictive model trained on a database where a quarter of addresses are outdated, email addresses are missing for the majority of records, and duplicate constituents have never been merged is not going to perform like a model trained on well-maintained data.

It will still produce scores. They will look authoritative. A meaningful percentage of them will be wrong — not because the model failed, but because it learned from broken information.

Garbage in, garbage out is not a clever phrase. It is an engineering reality. The organizations that get the best results from predictive modeling are the ones that treated data quality as a prerequisite rather than an afterthought.

The model is the engine. Clean data is the fuel. You can have the most sophisticated predictive model on the market, and if your database is in disarray, you have built a Ferrari and filled the tank with sand.

The Case for Better Tools

Here is what this means practically for anyone leading or staffing a development operation.

Reports are retrospective — use them to understand what has happened. Pivot tables are descriptive — use them to segment and analyze populations you already know. RFM is a useful triage tool — apply it where you need a manageable framework for basic segmentation decisions, and be clear-eyed about what it cannot see. None of these tools, alone or in combination, can score 50,000 records for giving likelihood.

None of them can find the prospect you don't know you should know. None of them can simultaneously weigh forty variables, account for their interactions, incorporate external data about what your donors give to other organizations, and rank every individual record in your database by the probability of a gift.

That is what a predictive model does. And at any scale above a handful of relationships, it is not a luxury. It is the difference between working your database and working the *right* part of your database.

The tool problem is real and it is solvable. Better analytical infrastructure exists, is accessible, and is producing measurable results in organizations that have adopted it. Ward and Pierce ran their campaigns on index cards and pencils.

The discipline of their thinking — identify the best prospects, focus your effort, make the ask — was exactly right. The computational power available to them was a fraction of what sits in your CRM right now.

A predictive model is just that discipline, running at a scale they never could have imagined, surfacing the people your mission deserves to reach.

Grace Hopper would have had no patience for organizations still sailing by dead reckoning when better navigation exists.

The tools have changed. The job hasn't. It's time for the methods to catch up.

There is, however, a second problem. And it isn't about the tools at all.

06 TRAPS

"The eye sees only what the mind is prepared to comprehend." — Henri Bergson.

Tools Aren't Enough

The previous chapter made the case against the tools — reports, pivot tables, and RFM — with specificity. The tools have structural limitations. They were designed for simpler data environments.

At scale, they miss what matters most: the donors you don't know you should know, hiding in databases nobody has examined closely enough.

That case is accurate. And it is, ultimately, the easier problem to solve. Better tools exist. The path to adopting them is well-documented. Organizations that have done it are reporting results.

The harder problem is this: the sector already knows the tools aren't enough. It has known for years. And it hasn't moved. Roger Craver has spent more than fifty years inside the machinery of nonprofit fundraising — building programs, testing approaches, watching the sector navigate every wave of technology since the age of direct mail. When asked why nonprofits have been so slow to adopt data-driven decision making, he doesn't hedge. He names the problem directly: innumeracy.

"The nonprofit industry by and large is innumerate," Craver has said. "They do not understand numbers, which is somewhat ironic since they deal with money and figures."

Innumeracy. Not ignorance. Not lack of access to tools. A fundamental discomfort with numbers — with measurement, with metrics, with the discipline of letting quantitative evidence determine what works and what doesn't.

It is a condition that runs deep in the sector, from the frontline fundraiser who trusts instinct over a predictive score to the chief executive who can describe every donor relationship from memory but cannot tell you what the organization's first-year retention rate was last year. It is, Craver argues, both a skill problem and a mindset problem — and of the two, the mindset is harder to fix.

This chapter is about the culture that produces that condition. About why it exists, why it persists, and why it has become — in the age of AI — something the sector can no longer afford to treat as a personality quirk.

The Wisdom Trap

Every experienced fundraiser carries a version of the same story. The major donor they knew was ready before any model said so. The appeal they felt in their gut would outperform the one the data recommended — and it did. The prospect the algorithm ranked low who turned out to be the organization's most transformational gift.

These stories are real. The pattern recognition that produces them is real. The conviction that relationship experience generates irreplaceable insight is not wrong.

It becomes a trap the moment it stops being a complement to measurement and becomes a substitute for it.

The wisdom trap works like this: years of experience produce genuine pattern recognition. That pattern recognition produces results. Those results reinforce the belief that experience — not data, not models, not AI output — is the primary asset.

And that belief, unchecked, produces resistance to any tool or system that threatens to redistribute authority from the person who has been doing this for twenty years to the algorithm that has been running for twenty minutes.

The trap is not that the experienced fundraiser is wrong about their own knowledge. It is that organizations have always been better at *knowing* things than *measuring* things. And the gap between those two words is exactly where the wisdom trap lives.

Human memory is a selective instrument. It remembers successes more vividly than failures. It remembers the donors who were visible — who attended events, returned calls, sent thank-you notes — and systematically forgets the ones who were quiet.

It builds pattern recognition from a sample that is never representative of the full database, because the full database contains tens of thousands of people no one has ever met.

The model doesn't have those limitations. It finds the donor who has been quietly giving $250 every year for eighteen years, lives three states away, has never attended an event, and would score as a planned giving prospect with a high likelihood of a six-figure bequest — the person that a gift officer's memory, working from its natural sample, would never surface. Those aren't competing capabilities. They are sequential ones.

Gut instinct tells you how to work a relationship. The model tells you which relationships to work. Organizations that treat them as adversaries are leaving the most undiscovered potential in their files.

Lyman Pierce understood this dynamic nearly a century ago. He wrote that the greatest obstacle to a successful campaign was not the difficulty of the goal or the condition of the economy.

It was the organization that had done it before — the one that attributed its prior success to its own methods rather than asking what combination of methods, market conditions, timing, and donor base had produced the result, and whether any of those factors had changed.

The wisdom trap, in Pierce's framing, is the organization that raises $10 million and concludes it knows exactly why — rather than doing the harder work of understanding which elements of the campaign were replicable and which ones were luck. Success, Pierce observed, is the enemy of learning when it is attributed to the wrong causes.

Why the Trap Closes at Every Level

The wisdom trap is not limited to frontline fundraisers. It operates at every level of a nonprofit organization, and the mechanisms that spring it are different at different levels.

For the gift officer, it is usually about identity. The experienced major gift fundraiser has built a professional self-concept around relationship intelligence — the ability to read a donor, to know when to push and when to pull back, to sense the right moment for an ask.

Being told that an algorithm has scored a prospect higher than their instinct did is not just a data disagreement. It feels like a challenge to who they are and what they've spent a career developing. The resistance is not rational. It rarely is. It is the self-protective response of someone whose professional authority is being partially redistributed to a machine.

For the executive director, the trap is usually structural. Most nonprofit CEOs came up through program delivery, communications, or fundraising — not analytics. They were promoted because of their ability to inspire, build relationships, articulate mission with conviction.

Numbers were never the primary currency of their advancement. And once in the executive seat, they are surrounded by a particular kind of reporting: boards that want impact stories, donors that want to hear about lives changed, staff that brings them the wins.

The data that would tell a different story — retention rates quietly declining, the major gift pipeline thinning, the planned giving program that has identified no new prospects in three years — tends not to make it into the room unless someone has specifically built the systems and culture to surface it.

There is also the problem of vanity metrics. The sector has developed a remarkable facility for measuring things that feel like progress without producing accountability. Social media followers. Event attendance. Email open rates. Number of grants submitted. These are not meaningless — but they are, in Craver's precise formulation, measures of activity rather than measures of impact.

Things that can be counted without the discomfort of being held accountable to an outcome. A fundraising program that tracks open rates but not revenue per donor has not yet decided to be honest with itself about performance.

The board compounds the problem. Most nonprofit boards are composed of community leaders, major donors, and subject matter experts — not data professionals. They are selected for their networks, their generosity, and their passion for the mission.

Many of them share the same discomfort with numerical analysis that runs through the staff. The board meeting that reviews a fundraising dashboard with the same rigor it applies to an audited financial statement is the exception, not the norm. Without board-level demand for measurement discipline, the executive team rarely feels the urgency to develop it.

The Culture Problem That Becomes the AI Problem

Here is where innumeracy becomes something more than a management inconvenience. It becomes a structural barrier to AI adoption — and more specifically, to AI trust.

Resistance to AI in the nonprofit sector is frequently framed as a values question. Concerns about bias. Worries about dehumanizing donor relationships. Philosophical discomfort with machines making decisions. These concerns are legitimate and deserve serious engagement — Chapter 4 addresses them directly.

But beneath many of them, if you look carefully, is something simpler and less philosophical: a professional culture that has never been fully comfortable with quantitative reasoning, now being asked to trust outputs produced by the most quantitative reasoning systems ever built.

An organization that doesn't track first-year donor retention is not going to trust an AI churn prediction model. An organization whose gift officers have never been required to document cultivation activity in a CRM is not going to trust an AI-generated portfolio recommendation that scores a donor they've never heard of as a top prospect. The distrust of AI is often, at its root, the distrust of measurement — and that distrust was there long before AI arrived.

This is the critical point: the path to AI trust runs directly through the discipline of measurement. Not because measurement and AI are technically inseparable — they're not.

But because organizations that have built the habit of using numbers to make decisions are already oriented toward the posture that AI requires. Willingness to let an output challenge a prior assumption. Comfort with probability rather than certainty. Institutional confidence to act on a model score even when gut instinct points elsewhere.

The gift officers who were most resistant to predictive models were, almost universally, the same ones who had never been asked to justify their portfolio decisions with data. The ones who adapted most readily were, just as universally, the ones whose organizations had already built cultures of testing — where trying something the model recommended and tracking the result was normal practice, not a threat to professional autonomy.

Beth Kanter — author and one of the sector's most effective advocates for data culture — flunked math and once feared measurement. She went on to write one of the most influential books on nonprofit measurement ever published.

Her reflection on the journey: "I first thought using data was going to be really straightforward. It's just a technical skill. But it's not just a technical skill. There is a cultural aspect to it and there is a leadership piece of it too."

The technical skill can be learned. The cultural aspect — the willingness to be accountable to evidence — has to be built. And it has to be built before AI arrives, not as a response to it. Organizations that try to deploy AI into cultures that have never normalized measurement will discover that the tool is the least of their problems.

What the Cultural Failure Actually Costs

It is worth being specific about the price, because the sector has a tendency to treat measurement gaps as minor operational inefficiencies rather than strategic failures. First-year donor retention has hovered around 30% for decades. For every ten new donors acquired, seven will not make a second gift. The cost of acquiring those seven is entirely unrecovered. The lifetime value they would have generated is entirely foregone.

This is not a small problem — it is the most expensive failure mode in nonprofit fundraising. And it persists, in large part, because the organizations experiencing it are not measuring the right things with enough rigor to feel the urgency of fixing it.

The same calculus applies to major gift portfolio management. The gift officer managing 125 relationships based on memory and instinct will, over time, distribute attention in ways that feel right but aren't optimized. They will spend more time with donors who are enjoyable to visit and less time with donors who are harder to reach.

They will follow up on warm conversations and let cool ones go quiet. They will, without meaning to, recreate the same portfolio they've always had — same geography, same professional background, same giving range — because instinct returns to familiar patterns. A culture that measures portfolio health catches this. A culture that doesn't enshrines it.

And then there is planned giving — the program where measurement failure may exact its highest price. The donors most likely to make a bequest are often the most invisible constituents in a nonprofit's database. Loyal, long-tenured, modestly-giving. They rarely raise their hand.

Identifying them has historically depended on gift officers who happened to know them, or on donors volunteering the information. AI-powered planned giving identification changes this entirely — but only if the database contains the data the model needs, and only if someone has been measuring the right things long enough to build a foundation the model can learn from.

The cost of this cultural failure is not abstract. It is the retention revenue never recovered. The major gifts left in the portfolio because no one trusted the score. The planned gifts never discovered because no one built the model, and no one built the model because no one had decided that measurement was the organization's responsibility rather than someone else's.

It accumulates quietly, year after year, in the gap between what the organization raised and what it could have raised if it had been willing to be accountable to evidence.

The Conversion That Isn't a Training Course

The cure for innumeracy is not a statistics course. That is worth stating plainly, because the most common institutional response to data capability gaps is training — skills workshops, dashboard tutorials, analytics certification programs. These have value. They are not the solution.

The solution is a decision. A leadership decision made before any tool is deployed, any training is scheduled, any AI platform is purchased: the decision that the organization will be accountable to quantitative evidence of its own performance.

That decision has specific behavioral signatures. The executive director who asks, in every fundraising review meeting, not just how much was raised but what the retention rate was — and expects a specific number in response.

The chief development officer who requires gift officers to document cultivation activity in the CRM because portfolio recommendations depend on that data being complete. The board that reviews a fundraising scorecard with the same attention it gives the financial audit. The gift officer who visits a prospect the model scored high even though she's never heard of them — and records the outcome so the model can learn.

Jann Schultz, who has built data cultures at multiple organizations, was direct about her own starting point: "I actually still have nightmares about flunking third-grade long division." Her success with measurement didn't come from becoming a mathematician. It came from deciding that understanding and acting on metrics was a professional requirement — not optional, not someone else's job, but hers.

She made the metrics visible. She put them on the walls. She worked them into every meeting. She found partners who could do the heavy analytical lifting and invested the time to understand what the results actually meant.

That is what the conversion from innumeracy looks like in practice. Not a transformation into data scientists. A decision that the numbers matter — and the consistent, visible behavior that follows from it.

For organizations beginning this work in the age of AI, three things are foundational.

Choose fewer metrics and take them seriously. The organizations that made the most progress in *Data Driven Nonprofits* were not the ones tracking the most indicators.

They were the ones that identified two or three metrics that genuinely measured what mattered — first-year retention, multi-year retention, revenue per donor — and built organizational accountability around them. Not a dashboard of forty-seven KPIs. The three numbers that, if they improved, would actually change the organization's trajectory.

Make data visible. When information is accessible, people use it. When it requires effort to retrieve, it gets ignored. The development director who posts fundraising metrics outside the break room, who starts every team meeting with a retention number rather than a dollar total, who makes data a natural part of daily conversation rather than a quarterly report — that person is building the habit that AI readiness requires.

Reward honest failure. The culture that punishes the test that didn't work will never build the testing habit. The culture that treats a negative result as information — as something that teaches where not to invest next time — creates the conditions for iterative improvement. AI produces probability outputs, not certainties.

The organization that can't tolerate the prospect scored 850 who turned out not to be a major gift will never trust the model. The organization that tracks the outcome, feeds it back, and adjusts the approach has understood what measurement actually means.

The Both/And That Resolves It

There is a version of this argument that lands wrong — the version that positions data and AI as adversaries of relationship and wisdom, as if becoming more numerically accountable requires becoming less human. That reading is wrong and worth correcting directly.

Roger Craver built some of the most relationship-intensive fundraising programs in the sector's history. He is also one of the most forceful voices for measurement discipline. Beth Kanter admits to math anxiety and has championed data culture for thirty years.

The gift officers at the University of South Dakota Foundation who resisted predictive models — and then watched a colleague close a principal gift from a prospect the model surfaced — didn't stop being relationship builders when they converted.

They became better ones. Because the model told them where to go, and the relationship got them there.

Ward and Pierce built the most systematic fundraising operation of their era. They developed formulas, tracked metrics, held daily accountability meetings, telegraphed results to headquarters every night at nine o'clock.

They were also, by every account, extraordinarily gifted relationship builders who understood the human dynamics of asking at a depth that few of their contemporaries matched. The art and the science were not in competition. The metrics told them where to apply the art.

That is still the right framing. Measurement doesn't replace judgment. It focuses it. AI doesn't replace relationship. It identifies where the relationship needs to go.

The wisdom trap closes when an experienced professional decides that what they know is sufficient — that experience, relationship history, and institutional memory are enough to navigate a landscape that is changing faster than any individual can track.

It opens again the moment that same professional decides their knowledge and measurement are not in competition but in partnership. That the data can tell them things their memory can't. That the model can find people their instinct would never surface.

That trusting a score is not an admission of professional inadequacy — it is the decision to be accountable to evidence in a sector that has, for too long, been accountable mostly to itself.

The nonprofit sector has always been better at knowing things than measuring things. The gap between those two words — knowing and measuring — is where donors lapse unnoticed, major gifts go unsolicited, and planned giving prospects sit undiscovered for years in databases no one has looked at closely enough.

Closing that gap doesn't require becoming a different kind of organization. It requires becoming a more honest one.
Three numbers. On the wall. In every meeting. Tracked every quarter. Acted on every year.

The tool problem is solved with better tools. The culture problem is solved with a decision. The sector has the tools. The decision is the one that's still waiting to be made.

07 PRINCIPLES

"Honesty is the foundation of good design." — Dieter Rams

Principles are Not Constraints

In 1976, Dieter Rams looked around at the world of consumer product design and did not like what he saw. Cheap materials. Confused aesthetics. Products that demanded attention without earning it.

His response was not a manifesto of rebellion. It was a list of principles — ten of them — that described what good design actually was, as opposed to what it was pretending to be.

Rams was a designer, not a data scientist. But his logic transfers cleanly. Principles are not constraints. They are clarifications. They describe the essential nature of a thing so that practitioners stop reinventing the wheel every time they encounter a new surface.

When I wrote *Data Driven Nonprofits* in 2016, I borrowed Rams' framework to define what good data experience looked like for nonprofits.

Ten principles. Each one a description of data at its best — accurate, accessible, helpful, understandable, measurable, honest, trusted, actionable, prescriptive, and valuable.

The list was useful then. But data experience in 2016 was still primarily a human problem. People collected data, managed data, and used data. The challenge was mostly cultural and organizational.

That's no longer the full picture. Artificial intelligence has entered the system — not as a passive consumer of data, but as an active agent that learns from it, reasons with it, and acts on it at speeds and scales that no human team can match.

Which means the stakes attached to these principles have changed. Dramatically.

When a human analyst works with bad data, they notice. They push back. They ask questions. They catch the outlier that doesn't look right.

When an AI system works with bad data, it doesn't hesitate. It processes the input, finds the pattern, and produces the output — with complete confidence and at enormous speed. The errors don't slow down. They compound.

Good data principles have always mattered. In the age of AI, they are the difference between an organization that amplifies its mission and one that amplifies its mistakes.

Why Principles Before Tools

The most common mistake nonprofits make with AI is the same mistake they made with data a decade ago: they start with the tool, not the problem.

A new AI platform arrives promising to transform fundraising, personalize communications, predict donor behavior, and automate prospecting — all at once, all immediately, all at a fraction of current costs. The sales deck is compelling. The demos are impressive. The signing bonus is hard to resist.

Six months later, the platform is running. The outputs are flowing. And something is wrong — not catastrophically wrong, but persistently wrong. The donor segments don't match what the team knows to be true.

The predicted gift amounts are off in ways that matter. The personalized communications feel generic. The prospect scores lead gift officers toward the wrong doors.

The tool isn't the problem. The data is the problem. And the data is the problem because no one ever established what good data was supposed to look like before they asked a machine to learn from it.

This is the sequence that works: principles first, then practices, then tools. Understand what good data is. Build the habits that produce it. Then — and only then — give that data to systems capable of learning from it.

The ten principles that follow are not new. They are the same principles that have always described good data experience.

What is new is their urgency, their precision requirements, and the consequences of ignoring them. AI does not forgive sloppy data. It scales it.

The Ten Principles of Good Data in the Age of AI

01. Good data is accurate

Accuracy has always been the foundation — the address that actually reaches the donor, the gift amount that matches the check, the date that reflects reality and not a default. In the pre-AI era, inaccurate data was costly and frustrating. It produced returned mail, wrong reports, and awkward donor conversations.

In the AI era, inaccurate data is actively dangerous. Machine learning models trained on inaccurate historical records will produce confident predictions built on a false foundation. Every accuracy failure that gets fed into a model doesn't just persist — it teaches the model something wrong.

The No Garbage Rule applies here with particular force: we do not feed garbage to systems that cannot tell garbage from truth. Accuracy is not a nice-to-have. It is the prerequisite for everything else.

02. Good data is accessible
Data that exists but cannot be reached is data that cannot be used — and in the age of AI, data that cannot be used by the system is data that might as well not exist. Siloed records, locked spreadsheets, systems that don't talk to each other — these were inefficiencies before. Now they are architectural failures that limit what AI can learn and therefore what it can do.

Accessibility means that the right data is available to the right systems in the right format at the right time. It means eliminating the rogue spreadsheets that live on individual computers, outside the CRM, outside the reach of any analytical tool. The amnesty program for hidden data isn't just good data hygiene anymore — it's preparation for AI deployment.

03. Good data is helpful
Data is helpful when it supports decision-making across a range of situations — not just the situation it was originally collected for. A donor's communication preferences are helpful for email decisions. They're also helpful for gift officer visit preparation. They're also helpful for planned giving outreach. Data collected for one purpose should be designed to serve many.

AI systems benefit from richly helpful data because they find connections across variables that human analysts miss. The more genuinely helpful each data point is — the more it describes something real and usable about a constituent — the more a learning system can do with it.

04. Good data is understandable

Data that no one can interpret is data that no one will trust — and data that is not trusted will not be used, no matter how sophisticated the system that produced it. Understandability is a design problem as much as a data problem. Field names that made sense to the developer who built the system five years ago may be impenetrable to the gift officer trying to use it today.

AI makes this more urgent, not less. When a model surfaces a recommendation — this prospect is ready for a major gift conversation; this lapsed donor is likely to reactivate — the people acting on that recommendation need to understand what drove it. Explainability isn't just a technical feature. It's a trust requirement. If your team can't understand the data, they won't trust the AI that learns from it.

05. Good data is measurable

What cannot be measured cannot be improved. This was true in Pierce's era — when he tracked daily campaign progress on thermometers posted in public — and it is true now. Measurable data enables comparison: across time periods, across segments, across gift levels, across programs. It creates the baseline against which everything else is evaluated.

For AI specifically, measurability is how you validate that the system is working. If you cannot measure the quality of your data going in, you cannot evaluate the quality of the predictions coming out. Organizations that skip measurement at the input stage are flying blind at the output stage.

06. Good data is honest
Honest data does not flatter. It does not suppress the donor who lapsed, inflate the retention rate, or hide the campaign that underperformed. It reflects what actually happened — including what failed — because systems that learn from sanitized data learn the wrong lessons.

There is a particular temptation, in organizations under fundraising pressure, to let optimism color the record. Goals marked as achieved when they weren't quite. Retention numbers calculated in ways that make them look better.

Segments defined to emphasize success. AI trained on dishonest data learns to be confidently wrong. Honest data, even uncomfortable data, is the only kind worth feeding to a learning system.

07. Good data is trusted
Trust is not automatically granted. It is earned through consistency, accuracy, and transparency over time. A fundraising team that trusts its data uses it. A team that doesn't trust its data ignores it — or worse, uses it selectively, applying it only when it confirms what they already believe.

In an AI-driven organization, trust in data translates directly into adoption of AI tools. Gift officers who don't trust the underlying constituent records will not trust AI-generated prospect briefings built on those records.

Development directors who've been burned by bad data before will not act on AI-surfaced recommendations without second-guessing every output. Data trust is the foundation of AI adoption. Build it before you build anything else.

08. Good data is actionable

Data that produces no action is data that produces no value. Actionability means that the data is structured, complete, and relevant enough that someone — or some system — can do something with it. It means that the constituent record contains enough information to drive the next step, whether that step is a call, a visit, a proposal, or a stewardship touchpoint.

Generative AI has made actionability more visible and more testable than ever. Ask a language model to produce a personalized donor briefing and it will use everything available in the record.

If the record is sparse, the briefing will be generic. If the record is rich, the briefing will be specific. The actionability of your data becomes immediately apparent when you ask AI to act on it.

09. Good data is prescriptive

Prescriptive data doesn't just describe the past — it points toward the future. It provides enough context, structure, and history that the right next action becomes clear. In the nonprofit data maturity model, prescriptive analytics sat at the top of the scale because it represented the highest and most difficult use of data. Most organizations hadn't reached it.

AI has changed the economics of prescriptive intelligence. Machine learning models can now identify the patterns that predict behavior — who is likely to upgrade, who is at risk of lapsing, who has the capacity and inclination for a planned gift — with precision and at scale previously available only to the largest and best-resourced institutions.

But prescriptive AI requires prescriptive data. The model can only point toward the future as reliably as the data it was trained on describes the past.

10. Good data is valuable

Value is the culmination of all the other principles. Data that is accurate, accessible, helpful, understandable, measurable, honest, trusted, actionable, and prescriptive is data worth having. It is data worth maintaining, worth investing in, worth protecting. It is data that compounds over time — becoming more valuable as the records grow richer, the relationships deepen, and the patterns become clearer.

In the age of AI, organizational data is not just an operational asset. It is a strategic one. The organization that has maintained clean, complete, well-designed constituent records over years and decades has built something that cannot be purchased or replicated quickly. It has the raw material for AI that actually works — AI that learns from a true and complete history rather than a partial and corrupted one. That is not a small advantage. It is a compounding one.

The Infrastructure Beneath the Intelligence

There is a phrase that has been circulating in technology circles long enough to become cliché: garbage in, garbage out. It captures something true but misses the deeper point. The problem is not merely that bad inputs produce bad outputs.

The problem is that AI systems process bad inputs with the same confidence and velocity as good ones. They do not slow down to question the data. They do not raise their hand and ask for clarification. They learn — efficiently and at scale — from whatever they are given.

This means that data health is no longer just a best practice. It is the infrastructure beneath the intelligence. Before any AI tool is deployed, the question must be asked and honestly answered: Is this data good enough for a machine to learn from? Not perfect — perfect is a standard no organization meets and AI does not require it. But accurate enough, complete enough, honest enough that the patterns the model finds are real patterns rather than artifacts of how the data was collected, managed, and distorted over time.

Dieter Rams observed that good design is as little design as possible — that the best designs strip away everything that is not essential and leave only what is. The same logic applies to data. Good data is not the most data. It is the right data, maintained with discipline, designed for use, and ready to serve whatever purpose the organization needs it to serve — including purposes that didn't exist when the data was first collected.

In 1905, Charles Ward and Lyman Pierce built their fundraising systems on index cards, typewritten records, and handwritten ledgers. The discipline they applied to those records — the insistence on accuracy, the systematic updating, the careful tracking of every pledge and payment — was the intelligence behind their intelligence. The tools were primitive. The principles were sound.

Today the tools are extraordinary. They can process millions of records in seconds, identify patterns invisible to human analysis, generate personalized communications at scale, and predict donor behavior with accuracy that would have seemed like magic a generation ago. But none of that changes what was true in 1905 and remains true now: the intelligence is only as good as the data behind it.

Get the data right. Everything else follows from that.

08 FORCES

"Changes aren't permanent, but change is." — Neil Peart

The One Constant is Change

The conversations happening inside nonprofit organizations about AI are mostly the right conversations. How do we use it? What does it cost? Where do we start? Who on staff needs to be trained? These are reasonable questions, and the chapters that follow address them directly.

But there is a different set of forces bearing down on the nonprofit sector that most of those internal conversations are missing entirely. Forces that are not internal at all. Forces that will reshape the sector not because nonprofit leaders decided to change, but because the world they operate in is changing around them.

Peart understood the distinction that most technology conversations collapse. The specific tools that arrive in any given decade are temporary — they get replaced, upgraded, superseded. But the underlying change they represent is permanent.

The sector that treats each technology wave as a temporary disruption to be weathered, rather than as evidence of a permanent shift in how people give, how funders decide, and how collective knowledge can be organized, will find itself perpetually behind a curve it never quite reaches.

Three forces in particular deserve honest attention before we go any further.

The donors are changing. The funders are changing. And the sector itself is standing at the edge of an opportunity it has never had before.

Consumer Behavior Is Donor Behavior

There is a phrase from *Data Driven Nonprofits* that has held up better than almost anything else in that book: *consumer behavior is donor behavior.*

It was true when we were talking about smartphones in 2016. In 2015, 68 percent of adult Americans had a smartphone. By the time that statistic was published, nonprofits were already behind the curve — their donors had moved to mobile and their giving pages hadn't.

Organizations that understood this adapted. Organizations that dismissed it watched their online giving underperform, year after year, while their donors increasingly conducted the rest of their financial lives on a four-inch screen.

The same pattern has played out with every major technology wave that preceded AI. Direct mail felt impersonal and transactional in the early 1970s — until Common Cause went from zero to 500,000 dues-paying members in a year and the entire sector followed. Online giving felt unsafe until it didn't, and now exceeds $60 billion annually in the United States alone. Social media felt frivolous until it became a primary channel for both donor acquisition and emergency campaigns. Mobile felt like a niche until it wasn't. Each time, the same arc: skepticism, then adoption, then dependence.

AI will follow the same arc. The only question is how far behind the sector will be when it arrives.

What makes this moment different is the speed. Previous technology waves moved fast. AI is moving faster. Donors are already interacting with AI in their daily lives — in their banks, their streaming services, their healthcare systems, their shopping experiences.

They are being trained, every day, to expect personalization, to expect immediate response, to expect that systems know their history and anticipate their needs. Those expectations do not stay in one compartment of their lives. They migrate. Consumer behavior is donor behavior. It always has been.

The donor who uses AI to help manage their investment portfolio, who has AI-assisted conversations with their insurance company, who receives health recommendations calibrated to their specific medical history — that donor is going to notice when the nonprofit they have supported for fifteen years sends them a generic year-end appeal addressed to "Dear Friend."

They are going to feel the gap between the relationship the AI-assisted commercial world offers them and the relationship their favorite causes provide.

Some of them are already noticing. The ones who don't raise the complaint directly — who simply give less, or give elsewhere, or stop giving without explanation — are the ones who will never appear in an exit survey.
Roger Craver named the herd dynamic that governs how the sector eventually responds to this kind of pressure. One organization, or a small group of organizations, does something dramatically better than their peers. The herd notices. The herd follows.

The technology that enabled the breakthrough becomes standard practice within a decade. It happened with direct mail. It happened with online giving. It is happening with AI now — just faster, and with less time for the organizations in the back of the herd to catch up.

The DAF Complication

There is a parallel shift in donor behavior that deserves its own section, because it is changing the structure of American philanthropy in ways that AI will accelerate — for better and worse.

Donor Advised Funds. DAFs.

The growth of DAFs over the past two decades has been one of the most significant and least understood structural changes in the sector. What appears in the Giving USA data as a surge in foundation giving is, in significant part, a reflection of DAF growth — donors moving money into philanthropic vehicles that sit between the donor and the ultimate charitable recipient.

From faith to function. From proximity to purpose. From direct giving to managed philanthropy. The rearchitecting of American philanthropy has been underway for years, and most nonprofits are still trying to understand what it means for them.

Here is what it means: The donor who funds a DAF has already made the tax decision. They have already committed to giving. What they have not decided is where. That decision — which organizations receive grants from their DAF, in what amounts, on what timeline — is increasingly being made with more research, more intentionality, and more analytical sophistication than the direct donation it replaced.

AI accelerates this. A donor managing a DAF has access to AI tools that can analyze organizational financial health, program effectiveness data, leadership quality, and comparative impact across organizations in the same cause area.

They can query AI assistants for grant recommendations, generate due diligence summaries, and track the performance of their philanthropic portfolio. The DAF donor empowered by AI is a more sophisticated grant-maker than most program officers were a decade ago.

This is not a threat to nonprofits with genuine impact and healthy data practices. It is a structural advantage for them.

The organization that measures outcomes rigorously, publishes data honestly, maintains transparent financial records, and can articulate its impact with specificity is the organization that surfaces favorably in an AI-assisted donor due diligence process.

The organization that relies on narrative and relationship and proximity rather than evidence is increasingly invisible to the donor whose giving decisions are being informed by AI.

The consumer behavior insight applies here too. Donors managing DAFs are not a separate species. They are the same donors, with more capital and more patience. What AI changes is the information environment in which they make decisions.

The nonprofits that understand this and invest accordingly — in outcome measurement, in data transparency, in the ability to answer "what does your work actually produce?" with specific and credible evidence — are building structural advantages that compound with every passing year.

The DAF wave is not cresting. It is building. And the AI wave is building behind it.

The Funder Transformation

Now consider the other side of the funding relationship. For most of the history of organized philanthropy, foundations and institutional funders held a significant information advantage over the organizations they funded. They set the priorities.

They designed the reporting requirements. They determined what counted as evidence of impact. Grant-seekers adapted to their frameworks, often at significant cost to programmatic coherence, because the alternative was not getting funded.

AI is beginning to shift that dynamic — and not necessarily in ways that favor the organizations applying for grants. Program officers at major foundations are already using AI to triage grant applications, synthesize organizational data, flag inconsistencies between stated outcomes and actual results, and compare proposals against each other and against the evidence base for specific interventions. What once required a team of analysts and weeks of review can now be done in hours.

The foundation that receives 2,000 applications for 50 grants is not going to read 2,000 applications — it never was. But AI changes what "not reading" means. It means systematic, data-driven triage that is faster, more consistent, and increasingly more sophisticated than the human triage it replaces.

The implications for nonprofits are direct and practical. The grant application that is well-written but poorly structured for AI parsing may perform worse than it would have in a purely human review process. The outcome data that exists in narrative form — buried in program reports, described in qualitative terms, never captured as comparable metrics — is invisible to an AI system looking for evidence of impact.

The organization whose prior grant performance is documented inconsistently will not benefit from the contextual judgment a program officer might apply. It will simply score lower.

This is not a conspiracy. It is a foreseeable consequence of a technology being applied to a function that has always required triage. The organizations that understand this — that treat their grant applications as data artifacts, not just narrative documents — will have advantages that compound with every application cycle.

But the funder transformation goes deeper than triage. It reaches into how foundations and government agencies define impact, how they think about portfolio strategy, and how they evaluate whether their funding is actually working.

The large foundations that have been building data and evaluation capacity over the past decade are doing so in anticipation of this moment.

They are building toward a future in which the organizations that receive their largest grants are not simply the ones with compelling stories and trusted relationships — though those will remain essential — but the ones that can demonstrate, with real data evaluated against the evidence base for their field, that their programs produce the outcomes they claim. AI-assisted grant-making is not a threat to effective organizations.

It is a threat to ineffective organizations that have been protected by information asymmetry and relationship capital.

That protection is eroding.

Government funders are further along than most nonprofit leaders realize. Federal agencies managing performance-based contracts are increasingly deploying AI to monitor compliance in real time — flagging underperformance before the reporting period ends rather than discovering it during audit. State agencies are building algorithmic screening into RFP evaluation.

The organizations that have been managing government contracts through relationship and reputation are going to find the game increasingly technical. Data quality is not optional in that environment. It is a survival requirement.

The both/and here is exactly right. The funder-facing AI environment rewards the same organizational investments that the donor-facing AI environment rewards: rigorous outcome measurement, honest data, transparent reporting, and the capacity to demonstrate impact with specificity.

The organization that builds these capabilities because its funders are starting to require them will find that those same capabilities make it significantly more compelling to sophisticated donors managing DAFs. The investments compound across both relationships simultaneously.

The Sector Opportunity Nobody Is Talking About

Here is the largest idea in this chapter — and the one that gets the least airtime at every conference where AI is on the agenda.

The nonprofit sector exists to solve problems that markets won't and governments can't. Poverty. Disease. Injustice. Educational inequity. Environmental destruction.

The accumulated weight of human need that neither the profit motive nor the political process addresses adequately. These are not small problems. They are the most complex, most entrenched, most consequential problems in human civilization.

And for most of the sector's history, the organizations addressing these problems have operated in conditions of profound isolation.

Each organization building its own knowledge base. Each organization running its own programs, often overlapping with neighboring organizations doing nearly identical work. Each organization reporting to funders in incompatible formats, capturing outcomes in inconsistent ways, measuring success against frameworks that nobody outside the organization can compare or aggregate.

The result is a sector that knows an enormous amount and shares remarkably little of it. That has learned, in thousands of organizations over decades, what works and what doesn't — and has encoded that knowledge in grant reports that sit in foundation archives, in case notes that age in program databases, in practitioner experience that walks out the door when a program director retires.

AI changes this. Not immediately. Not automatically. But potentially — with intention and investment — at a scale that nothing before it has made possible.

Consider what becomes possible when AI can synthesize the outcomes data from thousands of workforce development programs, identify which program elements produce durable employment versus which produce only completion certificates, and surface those patterns in a form that practitioners anywhere can access and apply. That is not a technology product. That is the sector finally learning from itself at the speed the problems demand.

Consider what becomes possible when housing stability programs share anonymized client outcome data into a common infrastructure, build AI models that identify the intervention sequences most likely to prevent eviction for specific population profiles, and make those models available to every housing organization in the country — including the small community organization in a rural county that could never afford to build such a model independently. That is the scale argument made real.

Consider what becomes possible when the grant reporting that foundations require — the outcome narratives, the financial documentation, the performance data — flows into a shared infrastructure that allows funders to see, in aggregate, what is actually working across their entire portfolio rather than evaluating each grant in isolation. That is a transformation in how organized philanthropy understands its own impact.

None of this requires any organization to surrender autonomy or competitive advantage in fundraising. It requires something harder: the willingness to treat the sector's collective knowledge as a shared asset rather than each organization's proprietary territory.

The sector has attempted versions of this before. Collective impact frameworks emerged a decade ago with exactly this logic — the recognition that isolated organizational excellence was insufficient to move entrenched system-level problems, and that coordination around shared measurement and shared outcomes could produce effects that no single organization could achieve. Some of those efforts succeeded.

Many stalled on exactly the challenges that AI can address: the cost and difficulty of aggregating incompatible data, the impossibility of synthesizing thousands of reports manually, the inability to detect patterns across a portfolio without analytical infrastructure that most organizations couldn't afford.

AI lowers those barriers dramatically. The cost of synthesis is no longer the bottleneck. The technical complexity of building shared models is no longer the exclusive province of organizations with data science teams. The infrastructure question — who builds and maintains the shared platform — remains real. But the analytical capability required to make shared infrastructure valuable has arrived.

The breakthrough that triggers the herd dynamic Craver described — one organization, or a small group, doing something dramatically better than their peers — has not yet happened at the sector-intelligence level. But the conditions for it have never been more favorable.

The organizations that build it, whether individual technology providers, sector intermediaries, or coalitions of foundations with the appetite for collective investment, will set off a chain reaction. The sector will follow. It always does.

The problems that nonprofits exist to solve are not getting simpler. The resources available to address them are not growing proportionally to the need. The case for doing more with the same — for learning faster, for sharing what works, for building on each other's evidence rather than starting from scratch — has never been stronger.

AI doesn't make that case. The problems make that case. AI provides, for the first time, a credible technical pathway to acting on it.

The Permanent Change

Peart wrote that changes aren't permanent, but change is. He was describing something that the nonprofit sector has proven true with every technology wave of the past fifty years. Direct mail changed. Online giving changed. Mobile changed. Each specific implementation evolved, was superseded, was replaced.

But the underlying shift in how people give — toward more channels, more immediacy, more personalization, more evidence of impact — did not reverse. It compounded.
The reasonable response to the forces described in this chapter is to adapt incrementally. Train staff on AI tools. Update grant reporting formats. Refresh the website to meet donor expectations shaped by AI-assisted consumer experiences.

These are reasonable responses. They will produce incremental improvements. They will not change the arc.

George Bernard Shaw observed that: "The reasonable man adapts himself to the world; the unreasonable one persists in trying to adapt the world to himself. Therefore all progress depends on the unreasonable man."

Shaw believed that progress depends on the unreasonable person — the one who persists in trying to adapt the world rather than simply adapting to it.

The unreasonable response to these three forces is to ask what the nonprofit sector could become if it decided to use this moment to address the conditions that have constrained it for generations.

Organizations operating in isolation, unable to learn from each other at scale. Grant-making opaque and relationship-dependent rather than evidence-informed. Donor relationships transactional rather than genuinely personal.

Collective knowledge locked in siloed systems rather than accessible to everyone doing the work.

None of those conditions are permanent. They are artifacts of technical limitations that AI is beginning to remove. What replaces them depends entirely on what the sector chooses to build.

Ward and Pierce did not set out to transform American philanthropy. They set out to build a YMCA building in Washington in 1905, and in the process of doing it systematically and accountably, they discovered a method that raised $350 million in the next eighteen years and reshaped how the sector operated for the following century.

The unreasonable possibility here is the same one it always is: that the disciplined application of available tools — by people who believe the mission is worth the effort — can produce results that adapt the world to the mission rather than adapt the mission to the world.

The changes described in this chapter are not coming. They are here. The donors are already different. The funders are already changing. The sector-level opportunity is already technically within reach.

The question is not whether these forces will reshape the nonprofit sector. They will. The question is whether the sector shapes them deliberately — or discovers, a decade from now, that it adapted to the wrong things while the permanent changes remade everything else around it.

09 START

You don't have to be great to start, but you have to start to be great." — Zig Ziglar

Ready, Steady, Go!
The most common question about AI in the nonprofit sector isn't about technology. It's about permission.

Permission to experiment. Permission to fail. Permission to spend time on something that might not work. Permission to be the person in the room who raises their hand and says: we should be doing this differently.

That instinct — to ask for permission before starting — is worth examining. Because in most organizations, the permission question is not the real question. The real question is harder and more honest: Are we actually ready?

Not ready in the sense of having the right platform licensed or the right vendor contracted. Ready in the sense that the previous chapters have established as the genuine prerequisite — the right data, the right culture, and the right clarity about what problem AI is being asked to solve.

Those three things determine whether AI creates value or decorates the organization with expensive, confident-looking errors.

This chapter is about getting ready. Not in theory. Not at some future date when everything is cleaner and the timing is better.

Now, with the organization you have, assessed honestly, moving forward in the right sequence.

The Sequence Is the Strategy

There is a version of AI adoption that skips preparation and goes straight to deployment. It is the most common version. A platform gets purchased. A pilot gets announced. Dashboards fill with outputs. And six months later, the outputs aren't changing anything that matters, because the foundation they were built on was never examined.

Pierce was direct about this in the context of campaigns. He called it the difference between organizations that launched and organizations that prepared to launch.

The campaign that begins with three months of quiet, systematic preparation — prospect research completed, lead gifts secured, volunteer structure in place, case for support tested — performs at a different level than the campaign that announces its goal before it understands its prospect pool. Same tools. Same volunteer base. Often similar budgets. Completely different outcomes, determined before the first public event.

AI adoption follows the same arc. The organizations that will look back in five years and recognize themselves as genuinely AI-driven are not the ones that moved fastest in 2025 and 2026.

They are the ones that asked the right questions first — about data quality, about cultural readiness, about what specific problems AI was actually being asked to solve — and built on honest answers.

That sequence is not cautious. It is strategic. There is a meaningful difference.

The Three Readiness Questions

Before any AI tool is deployed, three questions deserve honest answers. Not aspirational ones. Not the answers that sound best in a board presentation. The actual answers, as the organization exists today.

Is our data clean enough for a machine to learn from?

This is the foundational question. AI systems do not evaluate the quality of what they are given. They learn from it — with complete confidence and at substantial speed. A model trained on a donor database where a quarter of email addresses are invalid, duplicate records have never been systematically merged, and wealth screening data hasn't been refreshed in three years will produce outputs that reflect that reality. They will look authoritative. A meaningful number of them will be wrong.

The standard here is not perfection. No organization's data is perfect, and AI does not require it. The standard is honesty. Can the organization answer basic data health questions with specific numbers? What percentage of records have valid deliverable addresses? Valid email addresses? What is the estimated duplicate rate? When was the last National Change of Address run? When was the last deceased suppression?

If these questions cannot be answered with specificity, the data foundation is not understood well enough to know whether AI can be built on it.

The answer to the data question determines the sequence of investment. Organizations whose data is reasonably healthy can move toward AI applications with confidence.

Organizations whose data has significant gaps or quality problems should invest in data health first — not as a delay, but as the most important AI investment they can make. AI built on broken data does not fail gracefully. It fails confidently, producing wrong answers with apparent precision, and making the underlying problem harder to detect and harder to fix.

Does our culture support evidence-based decision making? Data health is a necessary condition for effective AI. It is not a sufficient one. The second readiness question goes deeper: does the organization have a culture where evidence informs decisions — or does it have a culture where data is produced for reporting purposes and instinct determines action?

This distinction is not subtle in practice. The organization with a genuine evidence-based culture is the one where a gift officer will adjust her portfolio prioritization because the model surfaced a prospect she hadn't been calling — not because someone required her to, but because she trusts the evidence and acts on it.

Where a development director reports retention rates accurately, even when they are disappointing, because the culture treats honest data as the starting point for improvement rather than a performance to be managed. Where leadership asks hard questions that can only be answered with data, and expects specific numbers in response.

The organization without that culture will deploy AI and produce dashboards full of scores, flags, and recommendations — none of which will significantly change what anyone does, because the underlying habit of acting on evidence has never been built.

The tools will run. The organization will not change. And leadership will eventually conclude, incorrectly, that AI doesn't work — when the actual lesson is that AI cannot build a measurement culture from scratch. It can only amplify one that already exists.

Chapter 3 established that the seven culture types from *Data Driven Nonprofits* apply directly to AI readiness. The question to ask now is simple and specific: which of those culture types describes this organization, and is the answer strong enough to support meaningful AI deployment? If the honest answer is no — if decisions still run primarily on instinct and relationship, if data health is addressed reactively, if analytics are produced for leadership on request rather than embedded in daily workflows — then cultural work comes before AI work. Not instead of it. Before it. The sequence matters.

Do we know what specific problem we are trying to solve? The third question is the one most organizations skip. They know they want AI. They know the conference sessions are full of organizations using it. They know the sector is moving and they don't want to be left behind. What they haven't done is define, with specificity, what they are asking AI to accomplish.

This matters because general AI adoption produces general results. Specific AI applications, designed to address specific problems with measurable outcomes, produce results that can be evaluated and built upon. An organization that says "we want to use AI to improve fundraising" has not defined a problem.

An organization that says "we lose approximately fifteen percent of our $500-to-$999 donors annually and cannot identify them in time to intervene before they lapse" has defined a problem that AI is well-suited to address, with a clear success criterion: does the intervention reduce the lapse rate?

Pierce built his campaigns around exactly this kind of specific, measurable objective. Not a vague ambition to raise more money — a specific goal, a specific prospect pool, a specific timeline, and a specific structure of accountability.

The specificity was not bureaucratic precision for its own sake. It was the mechanism that connected organizational effort to defined outcomes and made it possible to know whether the campaign was on track.

The same discipline applied to AI means defining the problem before selecting the tool. Which part of the fundraising program has the clearest gap between current performance and achievable performance? Where is the data foundation strong enough to support a model? Which application would produce results visible enough to build organizational confidence and cultural momentum? The specific answers to these questions — not the general ambition to be AI-driven — determine where to start.

What Readiness Actually Looks Like

Readiness is not a destination. It is a continuum, and most organizations sit somewhere along it rather than at either extreme. The task is not to achieve perfect readiness before moving — it is to understand where the organization actually sits and to take the next right step from there.

For organizations with strong data health, a genuine evidence culture, and a clearly defined problem: the next step is deployment. The foundation exists. AI will amplify what is already working. The risk of moving is lower than the risk of waiting.

For organizations with reasonable data health but an underdeveloped measurement culture: the next step is cultural. Not a training program or a dashboard initiative — a leadership decision. The executive director who starts every fundraising review meeting with a retention number, who asks questions that can only be answered with data, who visibly uses evidence to make decisions — that person is building the culture faster than any tool can. The AI deployment follows from that change, not the reverse.

For organizations with significant data quality gaps: the next step is the data itself. Specifically and practically: one data health audit, one baseline measure of a specific quality dimension, one process that improves it over a defined period. Address deliverability. Email validity. Duplicate rate. Pick one, measure it, and move it.

Not because the data needs to be perfect before AI can be considered — but because the organization that has never asked how healthy its data is cannot answer the first readiness question, and therefore cannot make an informed decision about anything that depends on the answer.

For organizations that haven't yet defined a specific problem: the next step is the problem definition. Not a technology committee. Not a vendor evaluation. A conversation — among development leadership, program leadership, and the people who know the data best — about where the organization is losing ground that it should be gaining, and whether AI is the right instrument for addressing it.

The Permission Question, Answered

Return to where this chapter began. The most common question about AI is about permission.

Here is the answer: the permission is not the issue. The preparation is.

The organizations that have earned the right to move fast on AI are the ones that did the slow work first — the data quality investment, the measurement culture, the honest assessment of where they actually are. For them, speed is appropriate. The foundation is solid. The deployment will amplify what works.

The organizations that haven't done that work don't need more permission. They need a clear-eyed account of what comes first. Not as a delay. As the actual work — the work that determines whether everything that follows is built on something real.

Pierce's eleven success factors for campaigns were not a checklist of things to complete before starting. They were a description of what organizations that succeed actually have in place — competent leadership, a compelling and clearly articulated case, an informed constituency, tested methods, the organizational discipline to execute.

You can launch a campaign without them. The campaigns that have done so and succeeded despite the missing foundation are not evidence that the foundation doesn't matter. They are the exception that proves the rule — and they are considerably outnumbered by the campaigns that launched without the foundation and fell short.

The same is true of AI. You can deploy it without readiness. Some organizations have gotten lucky. Most have produced expensive confirmation that the data quality problem they were already aware of also affects AI performance.

Start where you are. Assess honestly. Build what is missing. Deploy on a foundation that can support what you are building.

That is not the slow path to being AI-driven. That is the only path that actually gets you there.

What First Steps Actually Look Like

The readiness assessment points to the first move. Not a general first move — a specific one, determined by where the honest assessment puts the organization.

If the data foundation is the gap: conduct one structured data health audit this quarter. Choose a specific, measurable dimension — email address validity, address deliverability, duplicate record rate. Establish a baseline with a specific number. Build one process that moves that number over ninety days. Track it. Report it. Then choose the next dimension. Data health is not a project with an end date. It is a discipline that compounds, and the organization that starts the discipline today is not behind the organization that deployed an AI tool six months ago on broken data. It is ahead of it.

If measurement culture is the gap: choose three metrics that genuinely measure fundraising health — first-year retention, multi-year retention, and revenue per donor — and make them visible. Post them. Put them in every meeting. Ask about them before asking about dollars raised. The leader who does this consistently for ninety days has started building the culture that AI requires.

That is not a small thing. It is the specific behavior change that separates organizations that will benefit from AI from those that will purchase it and wonder why it didn't help.

If problem definition is the gap: schedule one conversation — not a committee, not a retreat — between the development director and whoever knows the data best. The question on the table: where is the clearest gap between what we are achieving and what we should be achieving, and is it specific enough to define a problem that AI could address? Come out of that conversation with one answer. Not a strategy. One specific, measurable problem. That is the starting point.

If the foundation is solid: move. The previous chapters have laid out what is working today, what is coming next, and what measurement looks like when AI is deployed well. The organizations that have done the preparation work don't need more analysis. They need a first deployment, a clear success criterion, and the discipline to measure honestly and adjust accordingly.

Starting from Nothing

There is a version of this chapter that only speaks to organizations that are already part of the way there — organizations with a CRM, a development team, some data infrastructure, a vague awareness that retention is a number worth tracking. That is most of this book's audience, and the readiness framework above is designed for them.

But there is another reader. The executive director of a $1.2 million human services organization whose donor records live in a spreadsheet that three people maintain inconsistently. The development director who was hired six months ago and inherited a database nobody has cleaned in five years, a stack of thank-you letters that went out three weeks late, and a board that measures fundraising success by whether the annual gala made money. The solo fundraiser at a community health clinic who knows the organization does important work and has no idea where its best donors actually are in the file. This section is for them.

Starting from nothing is not the same as being behind. It is a different starting point, with a different first move. The organizations that built genuine data and AI capability didn't begin with platforms and predictive models.

They began with decisions — about what to measure, what to record, and what to do with what they learned. Those decisions are available to any organization at any budget level. The sequence is the same. The tools at each stage are different.

Step One: Know what you have.
Before any investment in tools, systems, or AI applications, spend thirty days understanding the current state of the constituent data. Not fixing it — understanding it. How many records are in the system, whatever system that is? How many have a valid mailing address? A valid email address? When did each person last give, and how much? How many people have given more than once?

These questions can be answered from a spreadsheet. They do not require a CRM, a data analyst, or a vendor. They require someone to look at what exists and count it honestly. The number that comes out of this exercise is not a grade. It is a baseline. Every organization that has built a serious data practice started with a baseline, even if the baseline was embarrassing.

The organizations most resistant to this step are the ones that suspect the baseline will be bad and would rather not know. That instinct is understandable and expensive. You cannot improve what you haven't measured.

And you cannot make a credible case for investment in better tools — to a board, to a funder, to a skeptical executive — without a specific number that demonstrates the cost of the current state.

Step Two: Pick one system and use it consistently.
The enemy of data health is not bad data. It is data in multiple places that nobody trusts. The spreadsheet on the development director's laptop. The separate spreadsheet the executive director keeps for major donors. The events list that lives in a board member's email. The donor database that nobody updates because the updates don't seem to matter.

Consolidation doesn't require a sophisticated CRM. It requires a decision: one place, used consistently, by everyone who touches constituent data. For some organizations that is a basic nonprofit CRM — many offer free or deeply discounted tiers for small organizations. For others it is a single well-structured spreadsheet with clear ownership and consistent update practices.

The tool matters less than the discipline. The organization that maintains one system consistently for twelve months has better data than the organization that purchased a sophisticated platform and populated it inconsistently for three years.

One system. Updated within 48 hours of any constituent interaction. Accessible to everyone who needs it. That discipline, maintained for a year, produces something that has genuine value: a coherent record of who your donors are, when they gave, how much, and how they've engaged. That record is the foundation that every application described in this book is built on. It cannot be purchased. It has to be built, one entry at a time.

Step Three: Measure three numbers and act on them. Chapter 11 establishes the three numbers that matter most: first-year donor retention, multi-year donor retention, and revenue per donor.

These are not metrics for sophisticated organizations. They are the minimum viable scorecard for any organization that wants to understand whether its fundraising is working.

For the organization starting from nothing, these three numbers are often revelatory. Most organizations that have never tracked first-year retention discover, when they calculate it for the first time, that it is somewhere between 20 and 40 percent.

That means six or seven of every ten new donors acquired last year are already gone. That number, seen clearly for the first time, is not discouraging — it is motivating. It names a specific, fixable problem.

It creates urgency that no conference session or vendor pitch has ever produced. And it gives the organization a success criterion for everything that follows: is this intervention improving retention?

Calculate these three numbers from whatever data exists. Even an imperfect calculation from an inconsistent spreadsheet is more useful than no calculation at all, because it establishes the baseline and creates the organizational habit of asking the question. The first year the numbers are rough.

The second year they are more accurate. The third year they are the foundation for an honest conversation about performance. That conversation is what an evidence-based culture is built from.

Step Four: Find one AI tool that solves one real problem.
A note before going further: this book does not name specific AI vendors, platforms, or products. That omission is intentional. The tools available to nonprofit organizations are changing faster than any publication cycle can track. Platforms that were category leaders eighteen months ago have already been superseded.

Startups that raised significant funding to serve the nonprofit sector specifically have already consolidated, pivoted, or disappeared. Any vendor list published today would be partially wrong by the time this book reaches most readers, and confidently wrong within three years.

What doesn't change is the framework for evaluating any tool: does it address a specific, defined problem? Does it require data infrastructure the organization actually has? Can its results be measured against a criterion established before deployment? Those questions apply equally to whatever the market produces next year and the year after. The tools will keep changing. The questions won't.

With that framework in hand, the range of accessible starting points is wider than most small organizations realize. Free and low-cost AI writing tools can draft acknowledgment letters, cultivation emails, and appeal copy in minutes — freeing staff time for the relationship work that only humans can do.

AI-assisted wealth screening is available at price points accessible to organizations without enterprise budgets, and a single screening run on a constituent file can surface giving capacity information that changes how a gift officer prioritizes her next quarter.

Generative AI tools can synthesize program outcome data into donor impact reports, turning raw numbers into narratives that communicate without requiring a communications staff member to build them from scratch.

None of these applications require a CRM integration, a data science team, or a six-figure technology budget. They require a specific problem, a willingness to try something, and the discipline to measure whether it worked.

The key discipline for the first AI application is identical to what it is for the hundredth: define the success criterion before you start. Not "we want to save time" — how much time, on which task, measured how? Not "we want better donor communications" — what is the current open rate, response rate, or retention rate for this audience, and what would an improvement look like?

Without a defined success criterion, the first AI application produces an anecdote. With one, it produces evidence. Evidence is what builds the internal case for the next investment, and the one after that.

Step Five: Treat the learning as the asset.
The organization starting from nothing has one significant advantage over the organization that purchased an AI platform before it was ready: it has not yet made an expensive mistake that leadership will spend two years explaining.
The discipline of starting small, measuring honestly, and building from what actually works is not the consolation prize for organizations that can't afford to move faster. It is the methodology that the most effective data-driven organizations used to build capability that has lasted.

They didn't get there by purchasing the most sophisticated tool available. They got there by building a habit of using evidence to make decisions — starting with the smallest, most accessible evidence available and compounding from there.

Pierce's eleven success factors were not written for organizations that already had them all in place. They were written as a guide for organizations building toward them — a description of what a successful campaign looks like, backward-engineered into a preparation checklist.

The organization that reads those factors and says "we have none of these" is exactly the audience they were written for. The same is true here. Starting from nothing is not a disqualification. It is the most common starting point. The organizations that have built something significant all started somewhere that looked, at the time, inadequate to the ambition.

The data was a spreadsheet. Then a database. Then a source of genuine intelligence about donor relationships. The AI was a drafting tool someone tried on a Tuesday afternoon. Then a model that surfaced a major gift prospect nobody had called. Then an integrated capability that made the development program more effective at every level.

That arc does not begin with the platform. It begins with the decision to start — with what you have, from where you are, one honest measurement at a time.

The Both/And That Runs Through Everything

The argument of this chapter is not that organizations should wait. It is that preparation and movement are not opposites — they are sequential.

Ward and Pierce didn't launch campaigns the moment they arrived in a city. They spent weeks in what Pierce called the "unhurried period of preparation" — meeting community leaders, identifying the prospect pool, securing lead gifts, building the volunteer structure. And then they launched fast, with the confidence that the foundation would hold.

The unhurried period of preparation is not delay. It is the reason the campaign succeeds.

The same principle applies here. Assess the readiness honestly. Build what is missing. Then move with the confidence that the foundation will hold — because you built it on something real, not on the hope that the tool would be forgiving enough to compensate for what wasn't there.

The next chapter shows what moving actually looks like — not in theory, not at some well-resourced outlier institution, but in organizations of recognizable size doing recognizable work with the tools available right now.

Before you read it: answer the three readiness questions. Not aspirationally. Honestly.

The answer tells you exactly where to start.

10 NOW

"AI is whatever hasn't been done yet." — Larry Tesler

Living in the Now

There is a temptation, when talking about AI and nonprofits, to live entirely in the future tense. To describe capabilities that are coming, possibilities that are emerging, transformations that are almost here. It is more comfortable, somehow, to talk about what AI will do than to reckon honestly with what it is already doing — and what it means for organizations that haven't started yet.

This chapter is about the present tense. Not what AI might do for nonprofit fundraising. What it is doing right now, in organizations that have done the work, built the foundation, and deployed these tools with discipline and intention.

The applications described here are not experimental. They are not the province of large research universities with nine-figure endowments and armies of data scientists. They are being used today by organizations of varying sizes across every cause area.

The technology is accessible. The question is whether your organization is ready to use it.

Readiness, as always, begins with data.

Predictive Analytics for Major Gift Identification

This is where AI has already made its most measurable mark on the fundraising profession, and it's worth understanding precisely why.

For decades, prospect researchers relied on ZIP code-based wealth indicators — a useful but blunt instrument. The problem with ZIP codes is that they aggregate. They tell you the average profile of a neighborhood, not the specific profile of the person sitting in the house on the corner. A million-dollar home and the modest rental behind it share the same postal code.

Treating them as equivalent prospects wastes a gift officer's most valuable resource — time — on visits that were never going to produce major gifts, while the actual prospect down the street never gets a call.

AI changed that equation fundamentally. Today's predictive models don't just rank the donors you already know. They surface the donors you should know — the look-alikes who share behavioral and financial characteristics with your current major donors but have never been assigned to a portfolio.

These models examine giving history, yes, but they also incorporate engagement patterns, event attendance, email behavior, external charitable giving data, wealth screening results, real estate holdings, board affiliations, and dozens of other variables simultaneously.

No human analyst can do that at scale. No report can produce it. No pivot table can approximate it.

The distinction between prescriptive and predictive analytics matters here and is worth naming clearly. Prescriptive analytics helps an organization prioritize and rank order the people it already knows — those who attend events, respond to calls, participate in alumni associations. It is a better way of sorting a known list.

Predictive analytics goes further. It identifies the people in your database who you don't know but should know — the ones whose profile matches your best donors but who have never been on anyone's radar. It finds the diamond in the rough before the gift officer has ever heard the name.

Organizations running these models are reporting results that would have seemed improbable a generation ago. A 59% discovery visit conversion rate isn't magic. It's the mathematical consequence of directing limited resources toward the right people instead of guessing.

The University of South Dakota Foundation invested $15,000 in a modeling service and tracked $58.8 million in closed gifts against those model-predicted prospects. That number exists because someone tracked it — and because someone trusted the model enough to send gift officers to visit people they had never heard of.

That trust doesn't come automatically. It develops over time, as gift officers see the model produce hits, as leadership reinforces the expectation that data-informed visits are the expectation rather than the exception, and as the organization builds a culture where the model's outputs are treated as the beginning of the inquiry rather than a curiosity to be filed away.

The critical caveat deserves to be stated plainly: this only works when the underlying data is clean. A predictive model trained on a database where a quarter of addresses are unmailable doesn't fail gracefully — it fails confidently. It produces scores that look authoritative and outputs that look precise, while being built on a broken foundation.

The model is the engine. The data is the fuel. No engine runs well on contaminated fuel, and no model produces reliable outputs from unreliable data.

Lapse Prevention and Donor Retention Modeling

The best use of predictive analytics in fundraising isn't identifying your next major donor. It's identifying your current donors who are about to leave before they actually do. This distinction matters enormously, and it is where AI is delivering the most immediate and measurable value for organizations that have deployed it thoughtfully.

The economics of donor retention are straightforward and consistently underappreciated. Acquiring a new donor costs five to ten times more than retaining an existing one. First-year donor retention in the nonprofit sector has hovered around 30% for decades — meaning seven out of every ten new donors never make a second gift. The cost of acquiring those seven donors is entirely unrecovered.

The lifetime value they would have generated is entirely foregone. And that calculation doesn't even account for the planned giving pipeline: the donor who lapses after three years and never comes back was also a planned giving prospect who will never be discovered, and a bequest that will never materialize.

Donors don't leave all at once. They fade. They miss one appeal cycle, then another. Email open rates drop. Event attendance stops. The signals are there months before the final lapse — and in most organizations, nobody is watching for them because nobody has built the infrastructure to watch.

AI-powered retention models can detect the early signals of disengagement while there is still time to intervene. A donor whose email engagement has dropped steadily over six months. A mid-level donor who skipped their usual year-end gift for the first time in a decade. A long-tenured supporter whose giving frequency has shifted from three gifts per year to one.

Each of these patterns, in isolation, might escape notice. At scale, across a file of 50,000 donors, they are invisible without a model designed to find them.

The intervention doesn't have to be expensive or elaborate. A personal phone call. A handwritten note from a gift officer. A targeted email that simply says we noticed you've been with us for twelve years and we're grateful. What the model provides is the trigger — the knowledge that this donor, right now, is at risk — and the timing to act before the relationship ends rather than after.

Not every flagged donor needs the same intervention. Some need a personal touch. Some need a different communication channel. Some need a story that reconnects them to the mission outcome their giving has produced. The model identifies who is at risk. Human judgment determines what to do about it. That combination — machine precision and human relationship skill — is where the real value lives.

This is the kind of quiet, operational AI use that doesn't generate conference keynotes or vendor press releases. It generates retained donors, recovered relationships, and a revenue line that compounds quietly over years in ways that acquisition spending never can.

Planned Giving Prospect Identification

Half of all endowment funds ever created came from bequests. Read that again and then consider how most nonprofits market their planned giving programs — a page on the website, a legacy society brochure with a name that references heritage or legacy, and occasional mentions buried in the newsletter.

The gap between the magnitude of the opportunity and the investment most organizations make to capture it is one of the most persistent strategic failures in the sector.

The problem has always been identification. Planned giving prospects don't look like major donors. They don't necessarily have the wealth profiles that wealth screening catches. They are often modestly committed in terms of annual gift size — $100, $250, $500 — but extraordinary in terms of duration and loyalty. They give every year. They have given for decades.

They believe deeply in the mission. And they rarely raise their hand, because most of them have never been asked the right question or given a compelling reason to think of your organization in the context of their estate planning.

AI changes the identification equation entirely. A model trained on the profiles of known planned giving donors — giving longevity, gift frequency, recency, age and demographic indicators, estate planning signals, homeownership data, marital status, absence of major gift activity — can surface planned giving candidates at a scale and precision that no manual research process could approach. The multi-decade direct mail donor who has given $250 every year for thirty years and has never been approached about a bequest is not a difficult person to find with the right model.

They are hiding in plain sight in most nonprofit databases, invisible not because the data doesn't exist but because no one has built the system to look.

What AI cannot do is have the conversation. The cultivation of a planned giving prospect — building the relationship, understanding their motivations, making the case for a bequest, navigating the estate planning conversation with appropriate sensitivity — remains entirely human work. No algorithm closes a bequest. What the algorithm does is make sure the conversation gets started, with the right person, before the opportunity disappears.

Organizations that have deployed planned giving identification models consistently report the same finding: the prospect universe is far larger than they expected. They thought they knew who their planned giving candidates were. The model found many more. Most of those gifts are still undiscovered in databases across the sector. The bequest that arrives out of nowhere almost never comes from nowhere. The signals were there. AI can find them.

AI-Assisted Donor Communication and Content Personalization

This application gets more public attention than any other, and as a result generates both the most enthusiasm and the most anxiety. Both reactions are somewhat warranted, which is why this one requires the most careful thinking about how to deploy it well.

What AI is already doing effectively in this space is substantial. Drafting first versions of cultivation emails. Generating personalized acknowledgment language calibrated to a donor's giving history and engagement patterns. Segmenting audiences for communication based on behavioral data rather than simple recency and monetary tier. Testing message variations at a scale that would require armies of staff to accomplish manually. Adapting a single impact story for different donor segments — the same program outcome told differently to the first-time donor, the five-year annual fund supporter, and the major gift prospect who funded the specific initiative being reported on.

The equity argument for this capability deserves to be made explicitly. The mid-level donor segment — roughly the $1,000 to $10,000 range — has historically been under-served simply because there aren't enough hours in the day. A gift officer can manage 150 relationships with deep personal attention. The 3,000 donors in the mid-level range don't get that attention, because the math doesn't support it.

AI-powered personalization can give this segment a significantly better experience without proportional staff increases. It can reference a donor's specific giving history, connect their support to relevant program outcomes, vary tone and content based on channel and audience, and create communication that feels genuinely considered rather than mail-merged.

The risk — and it deserves an honest naming — is what might be called the efficiency trap. AI drafts the email. Nobody reviews it with real knowledge of the donor. The communication goes out faster and cheaper, and worse. The technology improved the production process and degraded the product. Personalization that was performed rather than genuine will eventually register with donors as hollow, and hollow communication erodes relationships rather than building them.

Organizations doing this well have built human review into the workflow. AI produces the first draft. A person who knows the donor, or who knows the audience segment, improves it. The output is better than what the human alone would have produced in the same time, and far better than what AI alone would have sent. That combination is the model worth replicating.

The fundamental principle Pierce identified in 1932 holds: content deployed without authentic conviction is not communication. It is noise. AI helps you produce more content faster. The human being in the workflow is what makes it worth reading.

Portfolio Management and Gift Officer Support

Gift officers spend a remarkable portion of their time doing things that are not relationship building. Preparing for donor visits. Pulling together giving history from multiple system sources. Researching recent news about a prospect's business interests, philanthropic activity, or family developments. Writing contact reports. Scheduling follow-ups. Updating cultivation plans. All of it is necessary. None of it requires the skills that make a great gift officer great.

The opportunity cost is significant. A gift officer who spends eight hours a week on research and administrative tasks has effectively sacrificed a full working day of relationship time every week. Over the course of a year, that is fifty days of visits, calls, and cultivation conversations that didn't happen. Over the course of a campaign, it is a fundamentally different pipeline.

AI is already being deployed to reduce this friction in measurable ways. Intelligent CRM integrations can aggregate prospect research from multiple sources, pull recent wealth screening data, surface relevant news and philanthropic activity, and present gift officers with a pre-visit brief that would have taken a full afternoon to assemble manually.

The briefing covers what you already know about the donor, what has happened publicly since your last contact, what questions you should be exploring in the next conversation, and what the model suggests about giving capacity and readiness.

Beyond visit preparation, AI is assisting with portfolio temperature monitoring — the ongoing work of tracking which relationships are warm, which are cooling, and which may be at risk of going cold. A model that analyzes giving history, engagement patterns, and contact records can identify donors who haven't been touched in too long, prospects whose wealth indicators suggest they're being asked for too little, and relationships showing early signs of disengagement.

The R-RG-RGF classification model — Resources, Resources plus Generosity, Resources plus Generosity plus Friendship — has always been the right framework for understanding a portfolio.

AI can apply it continuously, across the full file, rather than relying on a gift officer's memory to keep track of the temperature of 150 relationships simultaneously.

Contact report generation is another area where the friction reduction is tangible. A gift officer returning from a visit can speak their notes — what was discussed, what the donor expressed interest in, what the agreed-upon next steps are — and AI can draft a contact report from those notes that can be reviewed and filed in minutes rather than left in a queue for days.

This matters more than it sounds. Contact reports that aren't filed promptly are contact reports that don't fully capture the conversation, and institutional memory about donor relationships is one of the most fragile and valuable assets a development operation has.

The principle remains unchanged from Ward and Pierce: the ask happens across a table, not inside an algorithm. AI can tell you who to ask, when to ask, and what amount to anchor to. It cannot make the ask. What it can do is ensure that the person making the ask is better prepared, better informed, and better supported than they have ever been before.

Common Threads

The thread running through every one of these applications is the same. AI doesn't raise the money. It improves the quality of information available to the people who do. It frees capacity for relationship work by reducing the burden of research and administrative work. It surfaces what would otherwise be invisible — the at-risk donor, the planned giving prospect, the look-alike major donor candidate buried in the file.

And it scales attention to segments of the donor base that have historically been under-served because there weren't enough hours in the day.

The tools are here. The results are documented. The question is not whether AI works in nonprofit fundraising. That question has been answered. The question is whether your organization has the data foundation, the cultural readiness, and the leadership commitment to use it.

11 MEASURE

"If you torture the data long enough, it will confess to anything." — Ronald Coase

Measurement Foundations

The previous chapter described, in concrete terms, what is possible today. Predictive analytics surfacing planned giving prospects who have been hiding in your database for years. AI-generated briefings compressing hours of prospect research into minutes. Retention models flagging the lapsing donor six months before she's gone — while there is still time to call.

Those capabilities are real. They are deployed right now, in organizations across the sector, producing measurable results. The chapter made no promises it couldn't keep.

But it also made something clear that is easy to read past: every one of those applications depends on a foundation. Good data. A culture that acts on evidence. The discipline to let a model challenge an assumption rather than confirm one.

Without that foundation, the NOW chapter describes tools that underperform. The NEXT chapter — the emerging capabilities arriving in the next wave — describes tools that will be used by someone else.

This chapter is about that foundation. Specifically, about the discipline of measurement that separates AI that advances the mission from AI that generates the *appearance* of advancement.

It is not a glamorous subject. It does not generate conference keynotes. No vendor has ever built a sales deck around the argument that your measurement discipline is more important than their platform. But this is the chapter that determines whether everything in this book actually works — or whether your organization ends up with dashboards full of activity and outcomes that never changed.

The argument is simple and uncomfortable: AI generates more output than any technology in the history of the nonprofit sector. More communications. More scores. More briefings. More flags. More reports. More recommendations. All of that output creates a measurement problem the sector has never faced at this scale.

If organizations weren't disciplined about measuring what mattered before AI arrived, AI will make the problem exponentially worse. The volume of activity will increase dramatically. The clarity about whether any of it is working will decrease just as dramatically — unless someone decides in advance to measure the right things.

That decision — not a platform, not a model, not a vendor relationship — is what separates the organizations that will benefit from the AI era from the ones that will be buried in it.

The Dashboard Problem

Walk into most nonprofit development offices today and you will find dashboards. Boards with numbers on them. Reports that run every week. Emails from the CRM that generate automatically when a metric crosses a threshold.

More data than any development shop had available a decade ago, and almost certainly more than the team knows what to do with. And in most of those offices, first-year donor retention is still hovering around 30%.

The dashboards didn't fix it. The reports didn't fix it. The automated alerts didn't fix it. The problem isn't a shortage of numbers. The problem is that most of the numbers being tracked are what Roger Craver calls vanity metrics — things that can be counted without the discomfort of being held accountable to an outcome. Open rates. Click-through rates. Social media engagement. Email list size. Events attended. Unique visitors. Prospects flagged. Briefings generated.

These metrics feel like information. They look like progress. They fill slides nicely and they satisfy board members who want to see that things are happening. But they are, in Craver's precise formulation, things you can measure without doing anything about them. They describe activity. They do not describe results.

And the sector has been making peace with that distinction for decades, because measuring results requires honesty that measuring activity does not.

AI has supercharged the dashboard problem. Every AI tool generates metrics. Scores produced. Segments identified. Messages personalized. Recommendations made. Prospects surfaced. These are legitimate metrics — they describe what the tool did. They are not the metrics that tell you whether any of it mattered.

An organization that tracks the number of AI-generated donor briefings but not the visit rate, conversation quality, or cultivation advancement of the gift officers using those briefings has learned nothing useful about whether its AI investment is working. It has learned that the tool ran. That is not the same thing.

Coase's warning is not abstract here. It is operational. AI produces enough output that a determined organization — or a determined vendor — can arrange the numbers to tell almost any story. Response rates on AI-personalized emails look better than batch-and-blast. Prospect scores cluster more cleanly. Cultivation sequences complete at higher rates. All of that is true and none of it answers the question that matters: did more donors give more money and stay longer?

The data will confess to whatever you need it to confess to, if you ask only the questions that produce comfortable answers.

The organizations that will navigate the AI era well are the ones that resist the seduction of activity metrics and insist on the harder, more honest discipline of outcome metrics. They will track fewer numbers. They will post them on the wall. They will be accountable to them in every meeting.

And when the numbers don't move, they will ask why — not what the AI produced, but whether what the AI produced changed anything that matters.

The Nonprofit AI Paradox, Deepened

In the opening chapters of this book, the Nonprofit AI Paradox was defined plainly: AI has the potential to democratize insight and capability across the sector, but it is actually widening the divide between organizations that deploy it with discipline and those that deploy it with enthusiasm.

Organizations with good data, strong measurement culture, and competent leadership will compound their advantages. Organizations without those foundations will compound their disadvantages, because AI will be doing more — more efficiently, more quickly, more visibly — while producing less real progress.

Measurement is where the paradox becomes most acute. An organization with weak measurement culture that deploys AI will face a specific trap. The tool will produce outputs. Those outputs will look like progress.

Leadership will see dashboards with more numbers than they had before, and more of those numbers will be green. Staff will be busy acting on model recommendations, running AI-assisted cultivation sequences, generating personalized communications at scales previously impossible.

All of that activity will feel like advancement — and it will be nearly impossible to tell, from inside the activity, whether it is.

The trap is not malicious. It is natural. When you are doing more, it is easy to believe you are achieving more. When the tools are producing outputs continuously, it is easy to assume the outputs are producing results. When the dashboards are full and the meetings are busy and the workflows are running, it takes a deliberate act of discipline — an almost uncomfortable insistence on honest accounting — to stop and ask whether any of it is moving the only numbers that matter.

Donor retention. Dollars raised. Major gift conversion. Planned gift commitments.

Those are the numbers that measure whether fundraising is working. Not whether AI is working. Whether fundraising is working. If AI is deployed well, those numbers improve. If they don't improve, no volume of activity metrics changes that conclusion.

The organizations that will break out of the paradox are the ones honest enough to insist on that standard — before AI is deployed, not after.

What Pierce Measured

There is a reason Lyman Pierce required daily reports.
In the campaigns he ran between 1905 and 1923 — campaigns that raised more than $350 million in an era without CRMs, analytics platforms, or any form of digital communication — Pierce built accountability into the architecture of every campaign.

District Campaign Directors telegraphed results to headquarters every night at nine o'clock. Conference Campaign Directors sent totals by night letter. The numbers were not optional, not quarterly, not available on request. They were daily, mandatory, and public.

Pierce understood something that most nonprofit organizations still haven't fully internalized: measurement is not a reporting function. It is an accountability function. The difference matters enormously. Reporting tells you what happened. Accountability tells you whether you did what you said you were going to do — and exposes the gap when you didn't.

The daily telegraph was not about generating data. Pierce didn't collect those numbers to populate a dashboard or satisfy a board requirement. He collected them because the act of measuring daily, publicly, and with full visibility across the campaign team changed behavior.

Gift officers knew the numbers were coming. They knew their results would be visible. They knew the gap between projection and performance would be exposed without delay.

That accountability structure — the knowledge that the numbers mattered and that someone would act on them — is what drove the extraordinary performance those campaigns achieved.

Fast forward to today and the principle is identical, even if the telegraph has been replaced by Salesforce. The organizations that use AI most effectively aren't the ones with the most sophisticated models.

They're the ones with the clearest accountability structures — the ones where gift officers know that cultivation activity is tracked, that AI recommendations are recorded along with outcomes, that the model's suggestions are evaluated against actual results, and that the gap between predicted and actual performance is analyzed and fed back into the system.

That is measurement as accountability. Everything else is just counting.

The Three Numbers That Matter

In *Data Driven Nonprofits*, the argument was straightforward: every fundraiser needs three numbers. First-year retention. Multi-year retention. Lifetime value.

Those three metrics, tracked consistently and acted on relentlessly, tell an organization more about the health of its fundraising program than any dashboard of forty-seven KPIs ever will.

The argument has not changed — but one of the three numbers has. Donor lifetime value has been replaced here with Revenue Per Donor, and the reason is practical rather than theoretical: lifetime value is a projection that requires modeling assumptions many organizations don't have the data or capacity to make reliably.

Revenue Per Donor is a calculation any organization can make from the data it already has.

First-year retention is the canary in the fundraising mine. An organization that acquires ten new donors and retains three of them has a program problem — not an acquisition problem, not a communication problem, not an AI problem. A program problem. The first gift is an expression of interest. The second gift is an expression of relationship.

If 70% of donors never make a second gift, something is broken in the way the organization receives, acknowledges, and stewards that first expression of interest. No amount of AI-assisted personalization at the acquisition stage compensates for that failure at the retention stage.

Multi-year retention reveals the depth of the relationship program. Donors who give for three or more consecutive years have made a decision — not just about the organization's cause, but about the organization itself. They have decided it is trustworthy, competent, and worth a recurring commitment.

The organizations that build multi-year retention are the ones that steward gifts with intention, communicate with specificity, and treat the second and third gift with as much care as the first. AI can support all of those activities. But if multi-year retention isn't tracked, AI can't improve it, because no one is measuring whether it changed.

Revenue Per Donor is the number that connects fundraising activity to fundraising results. The calculation is straightforward: total revenue from contributions over a twelve-month period divided by the total number of unique donors in that same period.

No modeling required. No assumptions about future behavior. No projections that depend on data the organization may not have. Just a clear, honest answer to the question: on average, how much is each donor relationship generating?

That simplicity is the point. Donor lifetime value — the metric it replaces — is a projection. It estimates what a donor might be worth over the full course of the relationship, which requires assumptions about retention rates, giving trajectories, and future behavior that most organizations cannot make reliably.

The result is a number that sounds sophisticated and is frequently wrong. Revenue Per Donor is a calculation, not a projection. It is drawn from actual data, reflects actual performance, and can be calculated accurately from a spreadsheet.

An organization that tracks it year over year will see immediately whether the fundraising program is generating more from each donor relationship — or whether acquisition volume is masking a decline in individual donor value that nobody has named yet.

The ambition behind lifetime value — the conviction that donor relationships are long, not transactional, and worth sustained investment — is exactly right. Revenue Per Donor honors that ambition while keeping the measurement grounded in what actually happened rather than what a model predicts might happen.

If it is improving, the relationship program is working. If it is flat or declining while retention holds steady, ask amounts may need attention. If it is declining while retention also falls, the program has a compounding problem that demands immediate diagnosis.

These three numbers — first-year retention, multi-year retention, revenue per donor — are the test. If AI is working, they improve. If they are not improving, the right question is not what the AI is producing. The right question is why the relationship program isn't working.

The answer might involve AI. It might involve data quality. It might involve gift officer capacity or stewardship practice or communication strategy. But the three numbers are what tell you a problem exists and demand that you find out why.

Measuring the AI Itself

There is a second layer of measurement that the AI era requires — one that didn't exist when *Data Driven Nonprofits* was written, because the tools didn't exist. Organizations that deploy AI need to measure not just whether their fundraising outcomes are improving, but whether the AI is functioning as intended, producing reliable outputs, and operating within the boundaries of responsible use.

This is not primarily a technical problem. It is a governance problem.

Model performance degrades over time. The predictive model trained on two years of giving history will produce different outputs in year three — not because the model changed, but because donor behavior shifted and the training data no longer reflects the current population.

An organization that deploys a retention model and never evaluates its accuracy is not using AI. It is trusting AI blindly. Those are different things, and the distinction matters enormously when the model's outputs are directing where gift officers spend their time and where the organization invests its cultivation resources.

The governance question is: who is responsible for evaluating whether the model is still working? Not whether it is running — any technology team can verify that. Whether it is producing outputs that are accurate, fair, and aligned with the organization's values.

Whether the prospects it surfaces reflect the donor population the organization actually serves. Whether the communications it generates are consistent with the voice and relationship depth the organization has built with each donor.

These are not questions that answer themselves. They require someone to be responsible for asking them, on a defined schedule, with authority to act on the answers. In most nonprofit organizations today, that responsibility has not been assigned. The AI runs. The outputs go to gift officers or into communication queues.

No one is systematically evaluating whether the model is performing well, whether the outputs carry bias that disadvantages certain donors, or whether the personalization is genuine or performed.

Pierce's competent direction — his eleventh success factor, the principle that every campaign requires someone who knows whether it is working and can redirect resources when it isn't — applies directly here.

Not as a bureaucratic oversight function, but as the fundamental management responsibility that makes AI accountable to mission rather than accountable only to itself.

The Feedback Loop

The most important measurement discipline the AI era requires is one that most organizations haven't built: the systematic feedback of outcomes back into the models that generated the predictions.

AI learns from data. The predictive model produces a score. The gift officer acts on that score, visits the donor, records the outcome. The outcome — a gift, a declination, a conversation that revealed new information about the donor's interests or capacity — is the most valuable data point the model can receive. It tells the system whether the prediction was right, and in what ways it was wrong. It is how the model improves.

Most organizations are not feeding that loop. Some record outcomes in the CRM. Many don't, because the CRM is seen as an administrative burden rather than a strategic tool.

The model scores 10,000 prospects and the organization acts on 200 of them and records outcomes for maybe 50. The other 9,800 predictions go unevaluated. The model continues producing scores with no signal about whether those scores are moving closer to accuracy or drifting away from it.

The organizations that will compound their AI advantage over the next decade are the ones that treat outcome recording as a strategic requirement, not an administrative courtesy. Every model score that is acted on needs an outcome recorded.

Every AI-generated communication that goes to a donor needs an engagement result captured. Every cultivation sequence needs a conclusion logged — gift made, conversation had, relationship deepened, or donor disengaged. That data, accumulated over time with discipline, is what separates an AI that gets smarter from an AI that gets confident in its own errors.

This is, in the end, the same principle that Pierce built into his campaigns with daily telegraphed results. The feedback loop is the accountability structure. It is what converts output into learning. It is what makes measurement a discipline rather than a report

Honest Accounting

There is a version of AI adoption that produces dashboards, celebrates activity metrics, and declares victory before any of the outcomes have changed. It is the nonprofit equivalent of counting the number of times the campaign clock was updated without asking whether the campaign reached its goal. It is easy to do, socially comfortable, and professionally dangerous.

The honest accounting version looks different. It starts with three outcome metrics and refuses to be distracted by twenty activity metrics. It evaluates model performance on a defined schedule and acts on what the evaluation reveals.

It requires outcome recording as a non-negotiable operational standard, not a best practice someone might get to eventually. It connects every AI capability back to the question Pierce would have asked: is this campaign moving toward its goal, or merely moving?

That question is harder to answer than it sounds. Moving feels like progress. Dashboards full of green numbers feel like success. An AI that is running and generating outputs and keeping gift officers busy and producing reports that show continuous activity feels like a well-deployed AI. It might be.

The only way to know is to ask whether the three numbers are moving in the right direction.

First-year retention. Multi-year retention. Lifetime value.
If those numbers are improving, the AI is working. If they are not improving, the honest accounting tells you where to look. Not at the technology — at the program. Not at the model — at the relationship. Not at the activity — at the outcome.

This is what measurement as accountability looks like in the AI era. It is the same thing it looked like in Pierce's era, when the telegraph brought numbers in every night and someone at campaign headquarters had to decide what to do about them. The tools have changed beyond recognition. The discipline has not changed at all.

The organizations that measure honestly will know whether they are advancing the mission. The organizations that measure conveniently will know only that the AI is running.

That distinction — between honest measurement and convenient measurement — is the one that will define which organizations emerge from the AI era having built something durable, and which ones discover, too late, that their dashboards were full and their programs were stalled. Measure what matters. Post it on the wall. Act on it every quarter.

The tools will change. That discipline won't.

12 NEXT

"There is nothing new under the sun, but there are new suns."
— Octavia Butler

What Arrives Next

The most dangerous moment in any technology transition is not when the tools are unavailable. It is when the tools are arriving and the organization isn't ready for them.

The previous chapter described what AI is doing in nonprofit fundraising right now — predictive analytics surfacing major gift prospects, retention models catching lapsing donors before they disappear, AI-assisted briefings compressing hours of research into minutes. Those capabilities are real, deployed, and producing measurable results in organizations that built the right foundation before they reached for the tools.

This chapter is about what arrives next. Not in a decade. In the next 12 to 36 months. The capabilities described here are not speculative — they are the logical extension of what already exists, made possible by better models, better data infrastructure, and the hard organizational work that the early adopters in every sector are finishing right now.

But this chapter has a more important purpose than describing what's coming. It is about what you need to build today to be ready for it. Because every one of the capabilities described below requires something specific from the organizations that will use them well — a data foundation, a measurement culture, a workflow discipline — that takes time to develop.

The organizations that will extract full value from the next wave of AI are not the ones that move fastest when it arrives. They are the ones that are already prepared.

Octavia Butler was right. There is nothing new under the sun. The principles Ward and Pierce established in 1905 — identify the best prospects, cultivate with intention, ask at the right moment, steward the gift in ways that deepen the relationship — have not changed. They will not change. But there are new suns. And working by their light requires knowing what they illuminate.

What Gets Better First: Real-Time Donor Intelligence

The pre-visit briefing is already one of the highest-leverage applications of AI in major gift fundraising. Today's systems assemble what is known about a donor before a visit — giving history, wealth screening, recent news, cultivation notes — and compress hours of manual research into a document a gift officer can review in twenty minutes.

The next evolution removes the static limitation. Rather than a briefing assembled before a scheduled visit, continuous monitoring systems will track the information environment around each major donor relationship and surface updates the moment they become relevant. The donor's company announces an acquisition. Their daughter is appointed to the board of a peer organization.

A policy development directly related to their philanthropic priorities is published. Legislation they supported publicly passes into law. Each of these developments changes the conversation — and the gift officer who knows about them in the hours before a visit has a fundamentally different interaction than the one who walks in with last quarter's contact notes.

The downstream application is equally significant: AI-assisted conversation documentation that creates systematic institutional memory around donor relationships.

Today, when a gift officer leaves an organization, the accumulated understanding of how each relationship actually works — what the donor responds to, what they've resisted, what they've hinted at but never said directly — leaves with them.

The new gift officer inherits a folder of notes and a contact record, not a relationship. AI-assisted documentation changes that calculus.

The knowledge that has always walked out the door can be captured, structured, and transferred — not perfectly, but meaningfully. That is a different kind of organizational resilience than the sector has ever had.

What you need to be ready: Gift officers who document cultivation activity consistently, in a CRM that is maintained as a genuine record rather than an administrative obligation.

Contact reports filed within 48 hours. Cultivation notes that reflect what actually happened, not what sounded good in writing. The real-time intelligence system has nothing to work with if the baseline record is empty. The gift officer who has never filed a contact report is not a beneficiary of this capability. They are the reason it fails.

What Gets Better Next: Multi-Channel Journey Orchestration

The fragmentation problem in donor communications is one of the most persistent and expensive failures in development operations. The direct mail team optimizes within its lane. The email team optimizes within its lane. The major gifts team runs its own cultivation sequences. The digital team manages its own calendar.

A donor receives a mail appeal referencing a program she stopped supporting three years ago, an email the same week with a contradictory ask amount, and a gift officer call a month later from someone who doesn't know either communication went out. She experiences three separate organizations. Not one coherent relationship.

This is not a failure of intention. It is a failure of architecture. Fully integrated multi-channel orchestration addresses it at the architectural level.

AI systems will manage the full donor journey across every touchpoint simultaneously — adjusting email frequency based on engagement signals, triggering direct mail when behavioral patterns indicate a channel preference, suppressing automated communications when a personal conversation is active, routing a mid-level donor to a frontline fundraiser when behavioral signals suggest major gift potential is emerging.

The message, timing, channel, and content will be coordinated rather than managed in parallel silos that occasionally conflict and frequently ignore each other.

The compounding benefit is most visible at the mid-level. The donor in the $1,000 to $10,000 range has historically fallen into a gap — too significant for mass annual fund treatment, too numerous for genuine major gift attention.

Orchestrated AI engagement can give this segment a relationship experience that scales the practices of major gift stewardship without requiring the same staff-to-donor ratio. The moment when a mid-level donor's behavioral patterns suggest readiness for a major gift conversation can be identified automatically and routed to a gift officer before the opportunity cools.

What you need to be ready: A shared data infrastructure that every team in the development operation draws from and contributes to. Annual fund, major gifts, communications, and development operations all working from a unified view of each donor.

This is as much an organizational problem as a technical one. Orchestration technology built on top of siloed teams and disconnected systems will produce algorithmic fragmentation rather than coherent experience. The integration has to exist in the organization before the technology can reflect it.

What Gets Better After That: Individualized Impact Reporting at Scale

The annual impact report is a well-intentioned blunt instrument. Months of communications staff effort produce a document designed to speak to every constituency the organization has — major donors, annual fund supporters, foundation funders, board members, community stakeholders.

The result speaks especially well to none of them. The major donor who funded the capital campaign sees program statistics. The program funder who cares about outcomes sees facilities photographs. The twenty-year annual fund donor sees content designed for someone who discovered the organization last spring.

The best major gift stewardship has always addressed this problem manually. A seasoned gift officer writes individualized stewardship reports for top donors — connecting the specific gift to specific outcomes, in language that reflects what they know about the donor's interests and motivations, acknowledging the history of the relationship explicitly.

This work is among the most valuable stewardship a development operation produces. It is also among the most time-consuming, and it reaches a tiny fraction of the donor file.

AI will change the economics of individualized impact reporting fundamentally. Systems will generate stewardship narratives automatically, connecting a donor's specific giving history to program outcomes that align with their documented interests, in language calibrated to their engagement history and giving level.

The donor who has supported early childhood literacy for fifteen years receives an impact report organized around exactly those outcomes — with specific children, specific classrooms, specific measurements of change.

The donor who transitioned from annual fund to a named scholarship endowment three years ago receives a report that acknowledges that journey, updates them on the scholars their gift has supported, and invites a deeper conversation about what the endowment might become.

This is not a better version of the annual report. It is the replacement of a one-to-many broadcast with a one-to-one communication — the individualized stewardship that the best gift officers have always delivered to their most important relationships, now economically viable across a significantly larger segment of the donor file.

The work of reviewing and personalizing these AI-generated reports still belongs to humans who know the donors. What changes is the starting point: from a blank page to a complete, well-grounded draft that needs a human touch to become genuine. That is a different investment of staff time — and a better use of it.

What you need to be ready: Structured program outcome data connected to specific giving history at the constituent level. If the outcomes your programs produce exist only in narrative grant reports and annual report prose, there is nothing for an AI system to draw on.

The organizations that will benefit from individualized impact reporting are the ones that have built real outcome tracking infrastructure — where what changed for clients is captured in a form that can be connected to what donors funded. That infrastructure takes years to build. The time to start is not when the AI capability arrives. It is now.

What Gets Better Throughout: Dynamic Scoring Across the Full Donor Continuum

The R-RG-RGF classification model — Resources, Resources plus Generosity, Resources plus Generosity plus Friendship — is one of the most durable frameworks in major gift fundraising. Pierce described its elements nearly a century ago.

The logic remains exactly right: the prospect who has capacity, demonstrated philanthropic commitment elsewhere, and a genuine relationship with the organization and mission is fundamentally different from the prospect who has capacity alone. Working the full RGF profile has always been the basis of intelligent portfolio management.

The limitation of the model as historically applied is that it is static and manual. Scores are assigned based on a wealth screening and a research profile, reviewed annually if the organization is disciplined, and left in place long past their usefulness.

The donor who just sold a company doesn't get a new score until someone notices the announcement and updates it. The prospect who just made a transformational gift to a peer institution — a signal of both increased capacity and demonstrated generosity — doesn't have their RGF profile updated until a prospect researcher happens to come across the news.

The board member who recently stepped off a major civic role, now available for a deeper philanthropic conversation, doesn't surface in anyone's portfolio review.

Future AI will extend the RGF framework dynamically — updating scores continuously as new signals arrive, changes in financial circumstances recalibrating capacity scores immediately, new philanthropic commitments at peer institutions updating generosity signals in real time, shifts in engagement behavior flagging relationship temperature changes before they are visible to the naked eye.

More importantly, this dynamic scoring will operate across the full donor continuum rather than only at the major gift level where it has traditionally been concentrated. The first-year annual fund donor who shows engagement patterns consistent with a future mid-level supporter will be flagged early, while the relationship is young enough to shape deliberately.

The mid-level donor with emerging major gift indicators will be surfaced before the window closes — and before a peer organization's gift officer gets there first. The annual fund donor whose profile matches the planned giving prospect template will be identified for intentional outreach before they ever volunteer the information themselves.

The full file will be in motion. Not just the names at the top of the pyramid that everyone already knows. Every relationship at every level, evaluated continuously against the indicators that predict the next stage of the donor journey, flagged at the right moment for the right kind of human attention.

What you need to be ready: A CRM that is maintained as a living system rather than a historical archive. Consistent data entry practices across the full development team. Engagement data from every channel — email, events, volunteer activity, digital — captured in a form the system can read. Wealth screening refreshed on a regular cycle rather than run once and forgotten.

The dynamic scoring system has nothing to update if the underlying record hasn't been maintained. The organization that has been treating its constituent database as a mailing list will not benefit from a system designed to work with a relationship record.

The Preparation Principle

Pierce called it the unhurried period of preparation. It was the weeks before any campaign launched publicly — the quiet time when prospect research was completed, lead gifts were secured, the volunteer structure was assembled, and the case for support was tested against real audiences. Organizations that skipped this period and went straight to the public launch did not save time. They lost campaigns.

The same principle applies to what is described in this chapter. None of these capabilities are available to organizations that haven't done the foundational work. Real-time donor intelligence requires a maintained contact record.

Multi-channel orchestration requires organizational integration. Individualized impact reporting requires structured outcome data. Dynamic scoring requires a living constituent system.

Each of these prerequisites takes time to build — more time than the gap between now and when these capabilities become widely available.

The organizations that will be ready are the ones that treat the preparation as the work, not as a delay before the work begins.

What that looks like in practice is not complicated. It is the discipline described throughout this book — data health maintained with rigor, measurement culture built into daily workflows, CRM treated as a strategic asset rather than an administrative burden, outcome data captured in forms that can be used rather than formats that merely satisfy reporting requirements.

None of it is glamorous. All of it is the prerequisite for what comes next.

There is nothing new under the sun. The principles are the same ones that Ward and Pierce validated a hundred and twenty years ago. But there are new suns.

And the organizations illuminated by them will not be the ones that moved fastest when the tools arrived. They will be the ones that were ready when the tools arrived — because they had spent the intervening time building something worth illuminating.

Preparation without accountability, however, is just optimism with a plan. The next chapter addresses the other half of what readiness actually requires: the governance frameworks that ensure AI, once deployed, operates within the boundaries the organization's mission and values demand.

Not as a compliance exercise. As the structure that makes the preparation worth trusting.

13 GOVERNANCE

"The Analytical Engine has no power of originating anything. It can only do what we order it to perform." — Ada Lovelace

Where Governance Lives

There is a version of AI governance that organizations adopt to satisfy a checklist. A policy document appears. The board approves it. It gets filed in a shared drive somewhere between the travel reimbursement form and the facilities use agreement.

Nobody reads it again until something goes wrong.

That's not governance. That's paperwork with better intentions.

Real AI governance for nonprofit organizations is something harder and more valuable than a policy document. It is the ongoing practice of aligning how an organization uses intelligent technology with what the organization actually believes.

The two are not automatically the same. The gap between them — between capability and conviction, between what AI can do and what the organization should do — is exactly where governance lives.

The sector has seen this before. When data analytics arrived a decade ago, the organizations that handled it well weren't the ones with the most sophisticated privacy policies. They were the ones that had thought carefully about what their data meant, who it belonged to in a moral sense, and what obligations came with holding it.

The organizations that handled it poorly were the ones that treated data governance as a compliance exercise rather than a leadership practice.

AI raises the stakes considerably. The outputs move faster, operate at greater scale, and affect more decisions than any spreadsheet ever did. The governance question isn't whether to build a framework. It's whether to build one that actually guides behavior — or one that just looks good in a grant report.

What Happens When Governance Fails

Before describing what good governance looks like, it is worth being specific about what the absence of it actually produces. The nonprofit sector tends to learn more from other sectors' failures than from abstract warnings — and in the case of AI governance, there is no shortage of documented failures from adjacent fields that carry direct lessons for mission-driven organizations.

The first category of failure is algorithmic bias that produces discriminatory outcomes at scale — not because anyone intended discrimination, but because nobody looked for it before deployment.

In 2018, Amazon discovered that an AI recruiting tool it had been developing to screen resumes was systematically penalizing applications from women. The algorithm had been trained on ten years of the company's historical hiring data — a period during which Amazon, like most technology companies, had hired predominantly men.

The model learned from those patterns. It downgraded resumes containing the word "women's," flagged graduates of all-women's colleges as lower-ranked candidates, and consistently rated male applicants more favorably across comparable qualifications.

Nobody programmed it to do this. The bias was not intentional. It was the direct and predictable consequence of training a model on historical data that reflected a historical disparity — and then deploying it without adequate testing for differential outcomes. Amazon scrapped the tool when the bias was discovered internally. The decisions it had already shaped could not be recalled.

The lesson for nonprofits is not abstract. Any organization using AI to prioritize constituent outreach, identify major gift prospects, or allocate program services is building on historical data that reflects historical patterns — patterns that may include decades of underinvestment in certain communities, systematic gaps in who was asked to give, or disparate access to services that the organization is now using as a proxy for something else.

The algorithm will find those patterns and amplify them. Without deliberate testing for differential outcomes across population segments, the organization will not know this is happening until someone outside the organization finds it.

The second category of failure is data exposure through AI tools that organizations didn't adequately vet.

In 2023, Samsung engineers discovered that confidential source code had been inadvertently shared with an AI assistant as employees used the tool to help with debugging.

The sensitive information — proprietary code, meeting notes, internal hardware details — had been entered into a system whose terms of service allowed the provider to use inputs to train future models.

Samsung subsequently banned the use of generative AI tools on company devices. The data, once entered, could not be recalled.

The nonprofit equivalent of this failure happens whenever a staff member uses a free or consumer AI tool to draft communications that incorporate donor names, giving histories, or relationship details — or when program staff use AI tools to work with client data — without anyone having reviewed whether the tool's terms of service are consistent with the organization's privacy commitments and legal obligations.

Most free AI tools were not designed to handle sensitive constituent data. Most staff members who use them don't know that. And unlike a breach of an internal system, data shared with a third-party AI tool may not trigger the reporting obligations that would make the exposure visible — which means the organization can fail its constituents without ever knowing it has done so.

The third category of failure is AI-generated content that damages organizational reputation when it surfaces publicly.

In 2023, *Sports Illustrated* was found to have published articles under fabricated author names and bylines — human-sounding names with AI-generated headshot photographs — that were AI-generated and not adequately disclosed. When the practice was reported, the immediate reputational damage was significant and the long-term damage to reader trust was harder to calculate.

The publication had deployed AI content at scale without establishing standards for disclosure, human review, or quality control that would have prevented the specific failure.

The nonprofit sector's version of this failure is closer than most organizations realize. A fundraising communication that AI generated and no one who knows the donor reviewed before it was sent. An impact report that AI produced from program data that hadn't been verified. A grant application narrative that AI drafted and a staff member submitted without confirming that the specific claims were accurate.

Each of these represents content produced at AI speed with insufficient human review — and each represents a potential breach of trust with a donor, funder, or community partner whose relationship the organization cannot afford to lose.

None of these failures happened because the organizations involved were malicious. They happened because the organizations deployed AI capability faster than they developed governance frameworks to ensure that capability was used responsibly. The result in each case was predictable in retrospect and preventable in advance.

There is also a legal dimension that nonprofit leaders underestimate. AI use in employment decisions — including how staff are evaluated, promoted, or assigned work — is subject to anti-discrimination law in the same way any other employment practice is. AI use in decisions involving constituent data is subject to state and federal privacy regulations that are expanding rapidly.

The organization that uses AI to make decisions about individuals and has not thought carefully about those obligations is not protected by the fact that the algorithm made the recommendation. The legal exposure belongs to the organization that acted on it.

The governance argument, stated plainly: the absence of a real governance framework is not a neutral condition. It is a decision to allow AI to operate in the organization without accountability — and when something goes wrong, the absence of a framework is not evidence that the organization was being careful. It is evidence that the organization was not.

The Principles That Belong in Every Framework

The nonprofit sector doesn't need to invent AI principles from scratch. There is now a substantial and growing body of work — from technology researchers, government agencies, and international bodies — that has done the foundational thinking.

What the sector does need is to translate those principles into the specific context of mission-driven organizations with donor relationships, vulnerable constituencies, and limited resources.

Five principles belong in every nonprofit AI governance framework. Not because they appear on a list somewhere, but because each one maps directly onto something the sector already cares about.

Fairness and Inclusiveness

A predictive model that recommends which donors to contact, which clients to prioritize, or which constituents to reach is making decisions at scale.

If the data that trained the model was not representative — if it reflected historical patterns of who received services, who was asked to give, who was considered a good prospect — the model will encode those patterns and amplify them.

Bias doesn't enter AI systems through malice. It enters through history. And in the nonprofit sector, that history is sometimes uncomfortable.

Fairness in AI isn't about making every outcome equal. It's about ensuring that the tools aren't systematically disadvantaging people based on race, gender, geography, economic status, or any other characteristic that has nothing to do with the decision at hand.

For a healthcare organization using AI to identify grateful patients for major gift conversations, that means asking whether the model would perform differently across patient populations — and doing the work to find out before deploying it.

For a social services organization using AI to prioritize case management, it means auditing outputs for patterns that no one intended but that the data encoded anyway.

Inclusiveness extends the question from who gets equal treatment to who gets considered at all. AI systems built on historical data will systematically underperform for populations underrepresented in that data.

Organizations that serve diverse communities need to test their tools against the full range of people they serve — not just the ones most heavily represented in their records.

Privacy and Security

Nonprofit organizations hold some of the most sensitive data in the private sector. Donor giving histories. Client health records. Service utilization data. Financial circumstances documented in the course of major gift cultivation.

This data was given, in most cases, in a context of trust — donors trusted that their giving history would be used to advance the mission they supported, not fed into systems they were never told about.

Privacy in the age of AI isn't just about keeping data secure from external breaches — though that remains a baseline obligation.

It's about being honest with constituents about how their data is used, who can access it, what systems it trains, and what decisions it influences.

A donor who made a major gift fifteen years ago and whose wealth indicators are now feeding a machine learning model has a reasonable expectation of some explanation of what that means.

Security obligations compound when AI enters the picture. The attack surface expands. Data that flows into AI systems — and the outputs those systems produce — creates new vectors for exposure.

The nonprofit that stores donor data in a CRM it has properly secured but then pipes that data into a third-party AI tool without reviewing that vendor's data protection practices has not solved its security problem. It has outsourced it.

Transparency

The privacy policy last updated in 2015 is not adequate. The staff handbook that describes data processes as they existed before AI was part of the organizational infrastructure is not adequate.

Transparency doesn't require disclosing the technical architecture of every system to every constituent. It does require not deceiving people — or letting them operate under assumptions about organizational practice that are no longer accurate.

Transparency operates at three levels for nonprofit organizations:

With donors and constituents: honest disclosure about how information is used in AI-assisted decisions.

With staff: clear communication about which decisions are informed by AI outputs and what human review requirements apply.

With boards and leadership: regular, honest reporting on how AI systems are performing, what they're producing, and whether the results are what the organization expected.

The transparency principle has a particular application to AI-generated communications. When a development officer sends a "personal" note that was drafted by an AI tool and reviewed but not substantially rewritten by a human, there is a question worth answering: would the donor feel deceived if they knew?

The organizations that ask that question before deploying the practice are practicing transparency. The organizations that avoid asking it are not.

Accountability

Accountability is what happens when something goes wrong and someone owns it. In a traditional organizational structure, that's not complicated: a staff member makes a decision, the decision has a consequence, and someone is responsible for both.

AI complicates this chain in a specific way — when an algorithm recommends a course of action and a human acts on it without adequate review, accountability diffuses. The algorithm made the suggestion. The human didn't really evaluate it. Nobody feels responsible for the outcome.

That diffusion is not an accident. It's a foreseeable result of deploying AI without clear ownership. Governance frameworks that don't specify who is accountable for AI-influenced decisions — and what review processes exist before those decisions are made — are building diffusion into the system by design.

Human oversight must be a design principle, not an afterthought. The gift officer who reviews AI-generated prospect prioritizations before acting on them. The communications director who reads AI-drafted outreach before it goes out.

The development director who audits model outputs quarterly for patterns that suggest bias or error. These are accountability practices. They don't slow down the work as much as leaders sometimes fear. What they do is ensure that a human being remains responsible for what the organization does.

Accountability also requires a clear escalation path. When an AI system produces an output that concerns a staff member — a recommendation that seems wrong, a communication that doesn't feel right, a pattern that looks like it might reflect bias — there needs to be someone to tell and a process for review.

Organizations that lack this path will have concerned staff who stay quiet and problems that compound.

Reliability and Safety

An AI system that works brilliantly in a controlled pilot and fails unpredictably in production has not been adequately tested. Reliability means the tool does what it claims to do, consistently, across the full range of conditions in which the organization actually operates.

Safety means the consequences of failure — when failures occur — are bounded and manageable, not catastrophic.

For nonprofit organizations, the reliability question is closely connected to data quality. AI systems learn from the data they're given. A model trained on two years of clean donor data and deployed against a database with significant gaps, duplicates, and outdated information is going to produce unreliable outputs — not because the model is flawed, but because the inputs don't match what the model was built for.

Organizations that invest in data health before AI deployment are investing in reliability. The sequence matters. Safety in the nonprofit context means designing AI applications so that errors remain recoverable.

A prospect screening score that points a gift officer in the wrong direction is recoverable: the gift officer meets with the donor, the conversation reveals the mismatch, and the relationship continues. An automated communication system that sends the wrong message to ten thousand donors at once is significantly less recoverable.

High-volume, low-human-review applications require higher safety standards. The organization that treats them the same is taking risk it probably doesn't understand.

A Step-by-Step Plan for AI Governance Implementation

The principles above describe what AI governance needs to protect. What follows is how to build the infrastructure that actually protects it — practically, in sequence, in an organization that has real constraints and limited time.

This is not a theoretical framework. It is a working plan.

Step 1: Take Inventory Before You Build Policy

The most common mistake in AI governance is writing policy before conducting inventory. Organizations that do this produce frameworks that don't match reality — because they don't know what AI tools their staff are already using, what data is already flowing into third-party systems, or what decisions are already being shaped by algorithmic outputs.

Before drafting a single policy, conduct an honest inventory. What AI tools is the organization paying for? What AI features are embedded in commercial tools already in use — CRM systems, email platforms, document editors, communications tools?

What are staff members using on their own initiative, with or without organizational sanction? Where is donor or constituent data being processed by systems outside the organization's direct control?

The answers will be more extensive than most leaders expect. This is not a crisis — it's information. But it has to come first.

Step 2: Establish Clear Ownership
Governance requires an owner — a human being whose job includes making sure the governance framework is real, current, and followed. In larger organizations, this may be a dedicated role. In smaller ones, it is likely an addition to an existing portfolio.

The title matters less than the clarity: someone is responsible for AI governance, and that responsibility is documented, resourced, and reviewed.

This owner should not be exclusively from IT. AI governance for nonprofit organizations is fundamentally a leadership and values question, not a technical one.

The most important decisions about how AI tools get used involve mission alignment, constituent trust, and ethical judgment — none of which belong exclusively in a technology department.

The right owner typically sits in senior leadership, with clear channels to legal, IT, communications, and development.

Step 3: Define Your Approved Uses — and Your Prohibited Uses
With inventory complete and ownership established, the organization is ready to make actual decisions about AI use.

This means documenting — with specificity — what the organization has decided to do and what it has decided not to do.

Approved uses should include a documented rationale for each application. Not a comprehensive list of every possible AI application, but a clear record of the tools and workflows the organization has deliberately evaluated and chosen to adopt.

The rationale should explain what problem is being addressed, what the expected benefit is, and what review process occurred before adoption.

Prohibited uses deserve equal attention — and most governance frameworks underinvest in them. The prohibition list should be short and specific. A genuine ethical line, not a collection of vague concerns.

Examples worth including: using AI to generate donor communications that are represented as personally written by leadership when they have not been reviewed by that leadership; making individual decisions that materially affect constituents based on AI outputs with no human review; processing sensitive constituent data in third-party AI tools without appropriate data protection agreements.

Step 4: Build Data Governance That Precedes Deployment
Responsible AI requires knowing what data the system is learning from, where that data came from, whether it has been maintained with adequate quality, and whether its use is consistent with the organization's obligations to its constituents.

This is not a technical question delegated to a database administrator. It is a leadership question with real answers that senior staff need to be able to provide.

The data governance step means reviewing existing privacy commitments — in donor acknowledgment letters, on the organization's website, in grant agreements — and determining whether they accurately describe current data practice.

It means assessing data quality as a prerequisite for AI deployment, not an afterthought. And it means establishing clear rules about what constituent data may and may not be used in AI systems, including third-party tools.

Step 5: Establish Vendor Assessment Criteria
Most nonprofit organizations will not build their own AI systems. They will purchase them, license them, or encounter them embedded in platforms they already use.

That means a significant portion of the organization's AI governance happens through vendor relationships — and the organizations that fail to govern those relationships are outsourcing their values to companies whose interests are not identical to theirs.

Before adopting any AI tool that processes organizational data, establish and document the questions that will be asked of every vendor. What data does the tool use, and where does that data go? How is the model trained, and what testing has been done for bias? What security standards govern data storage and transmission? What are the terms under which the vendor can use organizational data to train future models?

What happens to organizational data if the relationship ends? These questions aren't hostile. Any reputable vendor will be able to answer them. Vendors who can't — or won't — have told the organization something important.

Step 6: Create Staff Guidelines That Reflect Reality
The staff handbook that prohibits AI use in general terms, in an era when AI features are embedded in nearly every commercial software platform, is not a governance document.

It is a fiction. Staff who encounter the prohibition will either comply and lose access to capabilities that could improve their work, or they will work around it and create exactly the undisclosed, unreviewed AI use that governance is supposed to prevent.

Effective staff guidelines start from a realistic assessment of how AI tools are actually used in the organization's daily work.

They define what responsible use looks like — when human review is required, what cannot be delegated to AI tools, what disclosure is expected when AI-generated content is used in donor-facing communications.

They provide examples that staff can actually apply, not principles so general they offer no guidance.

Staff guidelines also require training. A policy that people haven't read, in language they don't understand, addressing scenarios they haven't encountered, is not a governance mechanism. It's documentation. Training converts documentation into behavior.

Step 7: Design Human Oversight Into Every Workflow
Human oversight is not a check that happens after an AI system has already done its work. It is a design principle that shapes how the workflow operates from the beginning.

The gift officer who reviews an AI-generated prospect prioritization before acting on it needs a workflow that makes that review easy, expected, and documented — not an extra step that adds friction and gets skipped when things get busy.

Map the key AI-assisted workflows in the organization and design human review requirements into each one. Specify who reviews, what they are reviewing for, and what the escalation path is when something looks wrong. The goal is not to slow down AI systems — it's to ensure that a human being remains accountable for every consequential decision.

Step 8: Establish Review Cycles and Update Protocols
Responsible AI is not a one-time governance decision. It is a continuous practice. The AI tools an organization adopts will change. The data those tools operate on will change. The organizational context — staff, leadership, mission priorities — will change. The governance framework must keep pace.

Build regular review cycles into the governance structure from the beginning. Quarterly audits of AI outputs for patterns that suggest bias or error. Annual reviews of the approved use list to determine whether the tools are delivering what they promised. Annual updates to staff guidelines as new tools and new use cases emerge.

Board-level reporting that includes honest assessment of AI performance against organizational values — not just efficiency metrics.

The organization that reviews its AI governance annually and updates it honestly is practicing governance. The organization that files the policy and calls it done is practicing documentation.

Step 9: Engage the Board

AI governance belongs on the board agenda — not as a technology briefing, but as a fiduciary conversation. Boards have an obligation to ensure that organizational resources are deployed in service of mission and that organizational practices are consistent with stated values. AI deployment touches both.

Board engagement on AI governance doesn't require technical expertise. It requires the same judgment that boards apply to financial decisions, employment decisions, and strategic planning. Are we using these tools intentionally? Do we know what they're producing? Are we satisfied that our use of these tools is consistent with our obligations to the people we serve?

Leadership teams that don't bring AI to the board are making a governance decision by omission. The board that never asks about it is abdicating a responsibility it probably doesn't realize it has.

Step 10: Build a Culture Where Questions Are Welcomed

Every governance framework eventually confronts the gap between policy and practice. The policies are written. The training has happened. The vendor agreements are in place.

And then a staff member notices something that doesn't look right — a model recommendation that seems off, a communication that crossed a line, a data practice that doesn't match what the policy says.

What happens next is determined by culture, not policy. Organizations where concerns surface and get addressed learn and improve. Organizations where concerns get buried or dismissed amplify their problems until they become crises.

Build the culture deliberately. Name specific people whom staff should contact with AI-related concerns. Create a review process that takes those concerns seriously and communicates back. Reward the staff member who catches a model error.

Document the learning from every governance failure. The goal is an organization where the people closest to the work feel empowered to ask hard questions — because in AI governance, as in fundraising, the people closest to the work usually know what's actually happening.

The Governance Question Worth Asking

There is a version of AI governance that is primarily about protection — limiting liability, reducing risk, keeping the organization out of trouble. That version is not wrong. The risks are real and they deserve management.

The examples described earlier in this chapter — algorithmic bias encoding historical patterns and amplifying them at scale, data exposure through inadequately vetted tools, AI-generated content damaging reputations when it surfaces publicly — are not hypothetical. They are documented failures from organizations that did not ask the governance questions before deploying the capability.

But there is a better version: AI governance as an expression of organizational values. The framework that asks not just "what could go wrong?" but "what do we believe?" — and uses the answer to define both what the organization will do with AI and what it won't.

The nonprofit sector was built on trust. Donors trust that their gifts will be used as intended. Clients trust that their information will be protected. Communities trust that the organizations serving them have their interests at heart.

Every AI application an organization deploys either reinforces that trust or erodes it — depending not on the technology but on the choices made about how to use it.

Ada Lovelace, writing in 1843 about Charles Babbage's Analytical Engine — the mechanical precursor to the modern computer — understood something about machines that remains true of every AI system operating in the nonprofit sector today: they have no power of originating anything. They do what we order them to perform.

The accountability for what they perform, and for what it produces, belongs to the people who gave the orders — the leaders who chose the tools, the staff who configured them, the boards who never asked about them.

Governance is how those choices get made systematically instead of accidentally. It is how the organization ensures that the answer to "what do we believe?" stays visible in the daily operations of every tool, every workflow, and every decision that AI touches.

The tools will change. The obligation to deploy them honestly won't.

14 FUNDRAISING

"The genius of the great fundraising campaigns was not the dramatic ask — it was the invisible preparation."
— Lyman Pierce

How AI Creates Value

There is a temptation, when discussing AI and fundraising, to start with the technology. What tools are available. What platforms are emerging. What vendors are demonstrating at the conference.

That framing is backwards, and it has led a lot of organizations to spend significant money on systems that produce very little change in fundraising results.

The right place to start is the program. Annual giving. Mid-level giving. Major gifts. Principal gifts. Planned giving. Campaigns. These are not product categories. They are the organized expression of how a nonprofit builds relationships across its entire constituency — from the first-time $25 donor who responds to a direct mail appeal to the retired board member who ultimately leaves a transformational bequest.

The program is the strategy. AI is one of the tools that executes it.

This chapter walks the full program continuum and examines exactly where AI creates measurable value at each stage. It is not a chapter about technology.

It is a chapter about fundraising — informed by more than a century of proven practice, applied to the moment we are actually in.

The Annual Giving Program: Fixing the Leaky Bucket

First-year donor retention sits at approximately twenty-nine percent. Seven out of ten new donors to the average nonprofit never give a second time. Lapsed donor reactivation rates hover in single digits.

These numbers have not improved materially in decades, and they represent an enormous hidden cost — not just the sunk expense of acquisition, but the compounded lifetime value of relationships that were never built.

The annual giving program manager is managing a leaky bucket. The operational reality is relentless: multiple solicitation channels to coordinate, response rates that decline with each passing year, an aging direct mail file, and a digital program that acquires donors at lower average gift sizes than the channels it is supposed to supplement.

AI cannot fix the underlying problem if the underlying problem is a stewardship failure or a weak case for support. But it can do something that has historically been beyond the operational capacity of most annual fund teams: it can tell you exactly where the water is going, before the bucket runs dry.

Lapse prediction modeling is the most immediately valuable application. A model trained on the behavioral patterns of donors who lapsed in previous years — declining response rates, decreasing gift amounts, increasing time since last contact, disengagement from non-giving touchpoints — can flag at-risk donors months before they actually lapse. That window matters.

A personal call from a staff member to a donor who has given for seven consecutive years, made before the lapse rather than after, has a fundamentally different success rate than a reactivation appeal sent to someone who has already been gone for two years. The intervention cost is lower. The success rate is higher. The relationship is preserved rather than recovered.

Segmentation is the second major application, and here the contrast with current practice is sharpest. Most annual fund programs still segment their file primarily on recency, frequency, and monetary value — the RFM model that direct marketers borrowed from retail decades ago. RFM is not wrong; it is just incomplete.

A donor who gave $500 once three years ago looks identical to a donor who gave $500 once three years ago and also attended six events, opened every email, and referred two friends.

They are not identical prospects. One of them is being managed to the right outcome. The other is at serious risk of becoming a statistic. AI-powered segmentation can incorporate behavioral signals — engagement breadth, digital activity, event participation, channel preferences, social connection to other donors — and produce a far more nuanced prioritization than any RFM tier structure can achieve.

The third application is personalization at scale. The annual fund has always struggled with a fundamental tension: the donors who respond to mass communication are not the donors who should receive only mass communication.

AI can generate personalized acknowledgments calibrated to specific giving history, draft cultivation messages that reference a donor's actual program interests, and adapt a single impact story for thirty different audience segments — telling the same outcome differently to the first-time donor, the lapsed donor being reactivated, the multi-year loyalist being asked to upgrade, and the loyal $100 donor who, the model says, looks a great deal like a future mid-level supporter.

The economic argument is not subtle. Improving first-year retention by ten percentage points on a file of ten thousand donors produces a thousand relationships that would otherwise have been lost — relationships that, cultivated properly, will generate returns for years. AI is not the retention strategy. Genuine stewardship is the retention strategy. AI is what makes it possible to deliver genuine stewardship at a scale that manual effort cannot approach.

Mid-Level Giving: The Segment That Deserves Better

There is a band of donors in nearly every nonprofit's database who have been quietly under-served for years. They give somewhere between $1,000 and $10,000 annually.

They are too valuable to receive only mass communication, and too numerous to receive the personal attention that the major gifts program reserves for its portfolio. They exist in a gap — recognized in theory, under-resourced in practice.

This is not a new problem. But AI is, for the first time, making it solvable.

The math has always constrained this segment. A gift officer can carry 100 to 150 relationships with genuine personal attention. An organization with 3,000 donors in the mid-level range cannot staff its way to personal engagement at that volume.

What AI allows is a middle path: communication that is genuinely personalized — referencing actual giving history, connecting to specific program outcomes, varying in tone and content based on the individual donor's engagement profile — without requiring a gift officer hour for every piece.

The organizations doing this well are not automating their way to the appearance of relationship. They are using AI to produce first drafts that actual staff members review and refine — keeping the human in the loop not as a formality but as a quality control that makes the communication worth receiving.

The efficiency trap is real and worth naming here: AI drafts the email, nobody reads it carefully, it goes out generic regardless of what the template said, and a donor who has given loyally for a decade receives a message that could have been sent to anyone.

That outcome is worse than a thoughtful form letter, because it performs personalization without delivering it, and experienced donors can feel the difference.

The right model is AI as a first draft with human review as a quality control, not AI as an autonomous communication machine.

At mid-level, where the relationship is real but the portfolio is too large for traditional management, this model delivers something genuinely valuable: donors who feel known, who receive communications that reflect their history with the organization, and who are therefore far more likely to upgrade into the major gifts pipeline when the cultivation warrants it.

That upgrade path is where mid-level programs generate their highest return. AI can identify which mid-level donors are showing the behavioral signals of major gift readiness — increasing gift size, deepening engagement, capacity indicators that have shifted upward — and flag them for personal attention before they self-select out of the mass communication stream and into a relationship with a peer institution that got there first.

Major Gifts: Getting to the Table Better Prepared

Charles Ward and Lyman Pierce ran major gift programs on index cards. They rated prospects by capacity, tracked cultivation activity in handwritten notes, and managed solicitation timelines through sheer discipline and institutional memory.

The fundamental logic they used was exactly right: understand who has the resources, the demonstrated generosity, and the depth of relationship with your organization, and concentrate your finite capacity there. The R-RG-RGF framework — Resources, Resources plus Generosity, Resources plus Generosity plus Friendship — is not a modern invention.

It is a hundred years old. AI's contribution is not to change the logic. It is to apply it at a scale that index cards could never support.

The gift officer's core problem has not changed in a century: too many prospects, not enough time, and the constant pressure of competing demands on every hour. What has changed is how much of that time gets consumed by activities that do not require the skills that make a great gift officer great.

Research that used to take a full afternoon can now take twenty minutes. Before a major prospect visit, AI can aggregate publicly available information — philanthropic history, business background, board affiliations, public statements about causes and values, recent financial events — and synthesize it into a pre-visit briefing that positions the gift officer to have a substantially better first conversation.

The quality of a cultivation visit is directly connected to the depth of preparation that preceded it. AI does not make the visit. It makes the gift officer who walks through the door meaningfully better prepared for it.

Portfolio temperature monitoring is the second high-value application. A portfolio of 150 relationships is not a static list. It is a set of relationships at different temperatures — some warming, some cooling, some at risk of going cold without anyone having noticed.

AI can analyze giving history, engagement patterns, contact record frequency, and event participation to flag relationships that have been neglected too long, prospects whose capacity indicators suggest they have been systematically under-asked, and donors who are showing early signs of disengagement before those signs become a departure.

This is not replacing the gift officer's judgment about their donors. It is supplementing it with a systematic view that no human being can maintain simultaneously across 150 relationships.

Communication drafting deserves a direct word. The follow-up note after a cultivation visit. The proposal for a naming gift. The impact report for a donor whose named scholarship has been active for a decade. These are not mechanical tasks — they require tone, knowledge of the relationship, and genuine understanding of what matters to the person receiving them.

AI can produce strong first drafts that a gift officer then refines and personalizes. The goal is not to automate the communication. It is to make sure the communication actually happens, instead of sitting on a to-do list for three weeks because no one found the time to write it from scratch.

One thing must be said directly: AI cannot make the ask. The personal solicitation — face to face, from a person who believes in the cause and who knows the donor — remains the irreplaceable core of major gift fundraising. Ward and Pierce understood this.

They did not design systems to automate peer-to-peer solicitation. They designed systems to make it more systematic, better prepared, and more consistently executed.

AI is the modern infrastructure for that same discipline. Use it to get to the table better prepared. What happens at the table is still your job.

Principal Gifts and Planned Giving: The Hidden Prospect Universe

Half of all endowment funds ever created came from bequests. Let that sit for a moment. Not half of recent bequests. Half of all endowment funds — ever — trace back to a donor who chose to include an organization in their estate plan.

And yet most nonprofits operate planned giving programs that are materially underinvested relative to that return, marketed inconsistently, and staffed as an afterthought rather than a priority.

The traditional barriers to systematic planned giving identification have been real. Planned giving prospects don't self-identify. They don't wave a flag at a cultivation event or respond to a solicitation appeal in a way that makes their interest obvious.

The standard approach has been a combination of wealth screening, demographic guesswork — donors over a certain age, without obvious heirs, with long giving histories — and good fortune. Most planned giving officers will tell you that the gift that "came out of nowhere" almost never actually came from nowhere.

The signals were there for years. Nobody was looking for them systematically.

AI changes that equation. A model trained on the documented profiles of confirmed planned giving donors — giving longevity, gift frequency, recency, response to non-giving engagement, demographic characteristics, absence of major gift activity despite apparent capacity — can identify planned giving candidates across a large file with a precision that no manual review process could approach. The prospect universe that emerges is consistently larger than organizations expect.

Donors who have been receiving mass annual fund communications for thirty years. Donors who respond to every mailing but have never been contacted personally. Donors whose modest annual giving history obscures a capacity for a transformational gift at the end of their life. AI can find them. The bequest that arrives "out of nowhere" becomes, over time, the bequest that arrived because someone finally started looking.

Principal gift fundraising — the transformational gifts that define campaigns and reshape organizations — presents a different challenge. These gifts, typically seven figures and above at larger organizations, involve cultivation timelines measured in years, relationship depth that goes beyond professional engagement, and solicitation conversations that require extraordinary preparation and judgment.

AI's value here is primarily in the preparation: aggregating what is known about a prospect, surfacing what should be known, identifying alignment between the prospect's values and the organization's strategic priorities, and helping the gift officer develop the proposal and cultivation strategy that reflects genuine understanding of the donor's motivations.

The generational wealth transfer is not a background trend. It is the defining fundraising opportunity of the next two decades. As an estimated $84 trillion in assets passes from one generation to the next, the organizations that have built systematic planned giving programs — with AI-assisted identification, properly staffed cultivation, and genuine stewardship relationships — will capture an outsized share of what becomes available.

The organizations that are still treating planned giving as a brochure and a legacy society luncheon once a year will watch that opportunity pass them by entirely.

The mathematics of planned giving identification are simple and compelling. Most organizations have more planned giving prospects in their database than they have ever surfaced. AI can surface them.

What happens after that requires a human being who understands estate planning vehicles — bequests, charitable gift annuities, charitable remainder trusts, qualified charitable distributions, beneficiary designations — who can have a thoughtful, values-centered conversation with a donor about their legacy, and who has built enough of a relationship that the conversation is welcomed rather than deflected.

The algorithm identifies the prospect. The relationship determines the outcome.

Campaign Fundraising: The Discipline Ward and Pierce Built

Ward and Pierce would recognize a comprehensive campaign immediately. The quiet phase. The leadership gifts secured before the public launch. The gift table. The intensive period with daily accountability metrics. The volunteer structure organized around peer-to-peer solicitation. The public close.

These are not artifacts of a different era. They are the architecture of every successful major campaign conducted in the past century, because they are built on principles of human behavior that do not change.

What Ward and Pierce were doing with index cards and prospect rating committees was, at its core, a data exercise. They were building models — informal, manual, imprecise — of who had the capacity to give, who had demonstrated the generosity to give, and who had the relationship with the organization to give to this campaign.

They rated prospects. They assigned cultivation responsibility. They tracked progress against a gift table that told them whether the campaign was on track. The methodology was exactly right. The infrastructure was everything that 1905 had to offer.

Campaign planning today still involves informed judgment, historical giving data, feasibility study findings, and a substantial amount of educated guessing. The gift table is constructed before the campaign begins, based on the prospect pool as currently understood, and then treated largely as a fixed target. When gifts don't materialize at projected levels, the response is usually to work harder rather than to recalibrate the model.

AI changes both the pre-campaign analysis and the real-time management of an active campaign. Before the quiet phase begins, predictive modeling can assess the realistic major gift capacity of the prospect pool with a precision that no consultant's survey could approach — not the capacity that wealth screening suggests in isolation, but the capacity predicted by the combination of financial indicators, demonstrated generosity, depth of relationship, and cultivation readiness.

It can identify which prospects are ready for solicitation conversations and which still require cultivation work. It can flag the gap between current pipeline and campaign goal while there is still time to build the prospect universe rather than after the public launch when the gap becomes a crisis.

During the campaign itself, the gift table shifts from a static construct to a dynamic instrument. As commitments are secured, as new prospects are identified, as some donors move faster or slower than projected, the model updates continuously and surfaces the strategic adjustments campaign leadership needs to stay on course.

Which part of the pipeline is thinner than projected? Which prospects are ready for solicitation conversations that haven't been scheduled? Which leadership gifts are at risk of slipping past the window that would allow them to anchor the next tier of giving?

These questions have always been the right questions. AI can answer them in real time rather than at the quarterly campaign report.

The discipline Ward and Pierce pioneered was exactly right. The tools they had to execute it were a pencil, a prospect card, and the force of human conviction.

AI gives that discipline an engine that runs continuously, across the full prospect universe, updating with every new piece of information. The campaign model endures. What changes is the quality and speed of the intelligence behind it.

Advancement Services: The Back Office That Powers Everything

There is a staff function in most development operations that rarely appears in fundraising strategy discussions. It does not attend donor meetings. It does not make solicitations. It does not write cultivation letters or manage gift officer portfolios.

It processes gifts, manages data, produces reports, ensures compliance, reconciles financials, and maintains the infrastructure that every other part of the program depends on.

Advancement services — sometimes called development operations, or the back office — is the function that makes every visible aspect of fundraising work. And it is chronically underinvested, underrecognized, and underconnected to the technology investments that are transforming the programs it supports.

This is a problem, because the advancement services staff are the people who will determine whether AI creates value across the entire program or just in isolated pockets of it. If the data they maintain is incomplete, inconsistent, or outdated, the predictive models that rely on it will produce unreliable results.

If their gift processing workflows create errors or delays, the acknowledgment experience that shapes donor retention suffers. If their reporting infrastructure can't surface the metrics that matter, the program can't measure whether AI is actually working.

Gift processing has historically been a high-volume, error-prone, labor-intensive function. AI-assisted data entry validation — flagging inconsistencies at the point of entry, automatically applying gift coding rules, identifying likely duplicates before they enter the record — reduces the error rate at the front end of the process rather than requiring correction downstream.

Optical character recognition and intelligent document processing can handle the routine classification of gifts from paper checks, estate settlements, and complex gift instruments, freeing staff to focus on the exceptions and the judgment calls that genuinely require human expertise.

Data health — the persistent, ongoing work of keeping constituent records accurate, complete, and current — is where AI delivers some of its most unglamorous but highest-return value.

Address hygiene, email validation, phone number appending, deceased suppression, and lost constituent identification have historically required either significant vendor investment or staff hours that advancement services teams rarely have.

AI tools can automate much of this work, running continuous data quality checks rather than periodic batch processing, and flagging records that need human review rather than requiring staff to manually audit millions of rows.

The result is not just cleaner data — it is the prerequisite condition for every AI application that depends on data quality to produce reliable results.

Reporting and analytics is where the transformation of advancement services becomes most visible to the programs it serves. The shift from a back-office function that responds to ad hoc requests to a data service bureau that delivers proactive, self-service insights represents a fundamental change in how development operations contribute to fundraising strategy.

Florida State University Foundation's advancement services team, after consolidating multiple fundraising systems into a single platform, did not just improve data quality — they positioned themselves as the authoritative source of strategic intelligence for every fundraising program across the university.

That shift required technical investment, governance structure, and staff who understood both data management and fundraising well enough to translate between the two. AI makes that translation easier.

Natural language query interfaces allow gift officers and program managers to ask questions of the data in plain language rather than submitting reporting requests and waiting for a turnaround.

Automated dashboards surface the metrics that matter — retention rates, portfolio temperature, gift table progress, planned giving pipeline — without requiring a standing meeting to review a static report.

Prospect research and wealth screening, traditionally housed in advancement services, is perhaps the clearest example of how AI changes the capacity of a function that has always been constrained by staffing limits.

The work of building a research profile that used to take a full day — aggregating public records, philanthropic databases, news sources, real estate data, SEC filings, and board affiliations — can now take a fraction of that time.

What still requires the prospect researcher's expertise is the synthesis: deciding what the data means, identifying patterns that aren't obvious from the raw output, flagging discrepancies between what the wealth screening suggests and what the observed philanthropic behavior indicates.

The model handles the aggregation. The researcher applies the judgment. That combination — AI doing the volume work, humans doing the interpretive work — is the model that produces research profiles worth acting on.

The staff in advancement services are not background players. They are the keepers of the data that every other function depends on, and they are the people most positioned to understand whether that data is actually good enough to trust.

Organizations that invest in their advancement services teams — in training, tools, governance structures, and leadership that understands the strategic value of what they do — will find that AI performs better across every program.

Organizations that continue to treat advancement services as a cost center rather than a strategic asset will discover that their AI investments produce far less than expected, because the infrastructure those investments depend on was never built.

The Full Program, Connected

The fundraising program continuum — annual fund, mid-level, major gifts, principal gifts, planned giving, campaigns — has never been a series of independent silos, despite how most development offices are organized. Donors move across the continuum.

A $50 first-time donor who is cultivated properly becomes a $500 annual fund donor. A $500 annual fund donor who is identified early becomes a mid-level donor. A mid-level donor who is surfaced by a predictive model becomes a major gift prospect. A major gift donor who is stewarded deeply becomes a planned giving candidate. These transitions do not happen automatically. They happen because someone was paying attention at the right moment and had the right information to act on.

AI, at its best, makes the full continuum visible and keeps it in motion. The first-year annual fund donor who matches the profile of a future mid-level supporter gets flagged while the relationship is young enough to shape deliberately. The mid-level donor with emerging major gift indicators gets surfaced before a peer institution's gift officer gets there first. The loyal annual fund donor whose demographic and behavioral profile matches the planned giving prospect template gets identified before they ever volunteer the information. The major donor whose engagement has cooled gets personal attention before the relationship goes cold.

Ward and Pierce were working toward exactly this vision with the tools they had — index cards, prospect rating committees, systematic cultivation tracking. They understood that the data and the relationship were complements, not competitors. They understood that the disciplined application of what you know about donors, consistently applied across the full constituency, was the engine of every successful campaign they ran.

AI gives that discipline an infrastructure they could not have imagined. What does not change is the fundamental dynamic they identified more than a century ago: people give to causes they care about, presented by people they trust.

The AI-driven nonprofit does not replace that dynamic. It serves it — by making sure the right people hear about the right cause, that the relationships built are informed by the best available understanding of each donor's capacity, interests, and readiness, and that no prospect is overlooked simply because the file is too large and the staff too few to find them manually.

The tools will keep changing. The principles won't. And the organizations that understand the difference will raise more money, build deeper relationships, and advance their missions in ways that the technology alone — deployed without the principles behind it — never could.

15 PROGRAMS

"The purpose of a business is to create a customer. The purpose of a social business is to create change."
— Peter Drucker

The Larger Opportunity
Most of the conversation about AI in the nonprofit sector has been conducted inside the development office. How do we find better prospects? How do we personalize donor engagement? How do we predict which lapsed donors will reactivate?

These are legitimate questions, and earlier chapters have addressed them directly. But there is a larger opportunity hiding in plain sight — and it belongs not to the fundraisers, but to the program staff.

Consider what program people actually do. They design interventions and deliver services. They document outcomes and report to funders. They manage caseloads, supervise staff, navigate government contracts, respond to foundation reporting requirements, and measure whether the work is actually making a difference. They are, in the most literal sense, the people responsible for the mission. And they are operating — in most nonprofits — with tools that have not fundamentally changed in decades.

Here is the uncomfortable truth: AI may have more to offer the program side of a nonprofit than the fundraising side.

That is not a criticism of the chapters that came before. It is a recognition of where the bottlenecks actually are. Fundraising is already a professionalized function with sophisticated data infrastructure, established metrics, and decades of accumulated methodology.

Program delivery, in most nonprofits, is not. The data infrastructure is weaker. The measurement culture is thinner. The reporting burden is heavier and more manual. The potential for AI to create genuine leverage — not just efficiency — is correspondingly greater.

The argument of this chapter is straightforward: nonprofit organizations that confine AI to the development office will miss the deeper opportunity. The mission is not what happens when the fundraising succeeds. The mission is the whole point. Deploying AI in service of mission delivery is not a secondary use case. It is, ultimately, the primary one.

The Program Staff Problem

Ask a program director at most nonprofits how they spend their time. You will hear some version of the same answer.

They spend it managing people, responding to crises, attending meetings, and — in very large amounts — producing documentation. Grant reports. Outcome narratives. Compliance submissions. Funder site visit preparations. Government contract performance reports. Evaluation summaries. Logic model updates.

The documentation is not the mission. But the documentation is often what stands between the organization and its next round of funding. Program staff have become, by necessity, part-time grant writers, part-time data managers, and part-time compliance officers.

The actual work of designing better programs, supervising staff effectively, and deepening client relationships has been crowded out by the administrative burden of proving that the work is happening.

This is not a new observation.

Beth Kanter, who has spent over three decades helping nonprofits use technology to get better results, frames it precisely. "We have the human side and we have the technical side of measurement. They are kind of yin-and-yang. You need to have them perfectly balanced to really reap the benefits of continuous improvement."

The problem, she notes, is that nonprofits often see only one side — the tools, the metrics, the data — and neglect the human judgment required to make sense of what the data is actually saying.

AI addresses the documentation burden directly. It also addresses the measurement problem. And it has implications for program design, program delivery, and program revenue that most nonprofit leaders have not yet fully absorbed.

Responding to Funders: The Grant Report Problem

Foundation grants are the lifeblood of many nonprofit programs. They are also among the most time-consuming obligations a program staff can carry.

A single foundation relationship may require a letter of inquiry, a full proposal, a mid-year progress report, a final report, supplemental data submissions, and — if the funder conducts site visits — prepared materials for in-person review. Multiply that across a portfolio of ten, twenty, or thirty grants, and the math becomes punishing.

The typical grant report follows a predictable structure. The funder wants to know: What did you do? How many people did you serve? What changed for them? Did you spend the money as you said you would? What did you learn?

The answers are rarely mysterious — program staff know what happened. The challenge is assembling the narrative, pulling the numbers, reconciling the financial data, and producing a coherent document that reflects well on the organization within whatever deadline the funder has established.

This is precisely the kind of structured, information-intensive, deadline-driven task where AI excels.

A nonprofit with the right data infrastructure can use AI to draft grant reports from program data rather than producing them from scratch. The model draws from client service records, outcome tracking systems, attendance logs, and financial data to generate a first draft that captures what actually happened. Program staff review, edit, and add the human context and narrative specificity that a draft requires.

What once took three days to produce takes an afternoon. More importantly, the draft is better grounded in data than what a time-pressed program director working from memory might produce. The numbers are accurate because they came directly from the system of record.

The outcome narrative is consistent because it was generated from the same data being used in all other reporting. The funder receives a report that is more precise, more complete, and produced with less staff time than the version it replaced.

The downstream effect on funder relationships is real. Foundations that receive well-documented, data-rich, timely reports are more likely to renew grants.

They are also more likely to increase grant amounts and to make introductions to other funders. Grant reporting is not just a compliance obligation — it is a relationship maintenance tool.

Nonprofits that treat it as such, and that use AI to raise the quality of their reporting without burning out their program staff, will build funder relationships that compound over time.

Outcomes as Infrastructure

There is a deeper issue underneath the grant reporting problem, and it predates AI by decades. Funders increasingly demand proof of impact — not just activity, but change. How many individuals secured employment? How many students improved their reading scores? How many individuals stabilized in permanent housing after completing the program?

These outcome questions are harder to answer than activity questions, and most nonprofits are not measuring them with the rigor their funders now expect.

The transition from activity measurement to outcome measurement requires infrastructure. You need to know who you served. You need to track what happened to them. You need to connect the service they received to the change you're claiming produced. You need a data collection system that captures this information consistently and a staff culture that treats data entry as part of the work, not a bureaucratic add-on to it.

Most nonprofits have not built this infrastructure. This is the program-side version of the Nonprofit Data Paradox — they are generating data about clients and services continuously, but capturing it inconsistently, storing it in disconnected systems, and analyzing it rarely.

The data exists. The value hidden inside it does not emerge without deliberate investment in how it is collected, maintained, and used.

AI cannot substitute for this infrastructure. But it can accelerate the transition to it. Natural language processing tools can extract structured data from narrative case notes — turning the descriptions program staff write about client sessions into categorized, analyzable records.

Machine learning models can identify patterns in service delivery data that reveal which clients are most likely to achieve outcomes and which are at risk of disengaging. Automated dashboards can surface the three or four program metrics that matter most without requiring a standing meeting to review a manually assembled report.

The organizations that have done this work — that built real outcome tracking systems rather than performative ones — are positioned to use AI in ways that organizations with weak data infrastructure simply cannot. As we have argued throughout this book, the quality of AI output is bounded by the quality of the data it draws from. This is as true for programs as it is for fundraising.

Predicting Need Before It Becomes Crisis

Here is the program-side application of AI that most development-focused conversations entirely miss.

The clients most nonprofits serve are not uniformly distributed across the risk spectrum. Some are stable, engaged, and on track. Some are fragile, disengaged, and approaching a point where they will either reach out for help or drop out entirely.

The difference between those two groups is often visible in the data — in service attendance patterns, in the frequency of client contact, in case note content, in housing or employment stability indicators — before it becomes visible to a case manager carrying a caseload of fifty people.

A case manager carrying fifty clients cannot monitor all fifty with equal intensity. In practice, they triage — responding to the loudest signals, the most recent contacts, the clients who ask for help.

The clients who are quietly deteriorating but not yet in crisis often go unnoticed until the crisis arrives. By then, the intervention required is more intensive, more expensive, and less likely to succeed than an earlier one would have been.

Predictive models built on program data can change this. An AI system trained on historical client data — attendance patterns, service utilization, case note sentiment, demographic and contextual factors — can identify which currently stable clients show risk patterns consistent with disengagement or deterioration.

It can flag those clients for proactive outreach before the pattern becomes irreversible. The case manager's attention is directed not just to who is loudest, but to who most needs it.

This is not science fiction. Community health organizations, workforce development programs, and education nonprofits are already using versions of this approach.
The results are consistent with the theory: early identification enables earlier intervention, earlier intervention produces better outcomes, and better outcomes reduce the long-term cost of serving the same population.

The AI is doing what good supervisors have always tried to do — maintain awareness of the full caseload, not just the visible crises. What changes is the scale and reliability of that awareness.

The implications for funders are direct. A nonprofit that can demonstrate that its AI-assisted early identification system reduced client dropout rates by twenty percent is not making a vague claim about technology.

It is making a measurable claim about outcomes — the kind of claim that grant reports should contain and that funders increasingly require.

Personalizing Program Delivery

One-size-fits-all program design is a constraint of staffing and resources, not a reflection of what good program delivery actually looks like.

The workforce development program that offers the same curriculum to every participant regardless of their prior experience, learning pace, or employment barriers is not delivering the best possible service — it is delivering the most administratively manageable service. The distinction matters.

AI enables personalization at scale in ways that staffing ratios have always prevented. An educational nonprofit can use AI-powered adaptive learning systems that adjust curriculum difficulty and pacing based on individual student performance data. A workforce development organization can build AI-assisted case planning tools that recommend service sequences based on what has historically produced the best outcomes for clients with similar profiles. A mental health organization can use AI tools to match clients to specific intervention modalities based on presenting factors, rather than routing everyone through the same intake pathway.

The efficiency framing of these applications is real but insufficient. Yes, personalized delivery is more efficient — it doesn't waste a client's time on material they've already mastered, or route them to services they don't need. But the more important point is that personalized delivery is more effective.

Clients who receive services matched to their actual needs achieve better outcomes. And better outcomes are what the funders are asking for, what the board is measuring, and what the mission requires.

The staff capacity freed by AI-assisted program delivery does not simply disappear. It redirects toward the work that requires a human being — the complex cases, the relationship-dependent interventions, the clients whose situations don't fit the model. This is the both/and principle applied to programs:

AI handling the routine, humans handling the irreplaceable. Not one or the other. Both, working together, producing outcomes neither can achieve alone.

Fee-for-Service and Program Revenue

The revenue profile of the nonprofit sector surprises most people who encounter it for the first time. More than 80 percent of all nonprofit revenue in the United States comes from fee-for-service income and government grants and contracts.

Charitable contributions — the fundraising that occupies so much of the sector's attention and most of this book — represents roughly 14 percent of the total. For many nonprofits, the fee-for-service business is the business.

This matters for AI in a specific way. Fee-for-service programs are market-facing businesses operating inside nonprofit structures. They are subject to the same competitive pressures, operational constraints, and capacity limitations as any service business.

And they have the same opportunity to use AI to expand capacity, improve quality, and reduce unit costs without proportionally increasing headcount.

Consider a nonprofit vocational training organization that charges participants or government payers for certification programs. Its revenue is bounded by how many participants it can serve, how many instructors it can employ, and how many physical or virtual classrooms it can operate. AI-assisted instruction — adaptive learning platforms, AI-powered coaching tools, automated assessment and feedback — can expand the number of participants a single instructor effectively serves.

It can enable asynchronous delivery that removes scheduling constraints. It can provide the kind of individualized feedback that was previously only possible in small cohorts or one-on-one instruction.

The revenue implications are direct. More participants served with the same staffing level means higher revenue without proportional cost increases. And higher revenue from fee-for-service programs means more unrestricted resources available for the programs that can't generate revenue — the services to populations that cannot pay, the advocacy work, the community organizing, the mission activities that exist precisely because markets won't fund them.

AI in fee-for-service programs is not just a quality argument. It is a financial sustainability argument. Organizations that use AI to scale their earned revenue capacity free themselves from some portion of the grant dependence that creates perpetual instability.

That is a mission argument too — because organizations that are financially sustainable can make longer-term commitments to the communities they serve.

Responding to Government Contracts

Government grants and contracts represent the largest single revenue source for the nonprofit sector. They are also, by a significant margin, the most burdensome to administer.

Federal contracts in particular require compliance with detailed performance standards, quarterly and annual reporting against approved metrics, financial documentation at a level of granularity that most foundation grants do not approach, and — in some cases — multiple layers of audit and site review.

Program staff managing government-funded contracts spend enormous amounts of time on compliance. Not because the compliance is meaningful in every instance, but because the consequences of non-compliance — payment clawbacks, contract termination, debarment — are severe enough that organizations rationally over-invest in documentation to protect themselves.

AI can absorb a meaningful portion of this burden. Compliance reporting workflows that currently require manual data extraction, reconciliation, and narrative drafting can be automated to the point where a first draft of a quarterly performance report requires review and approval, not creation from scratch.

AI tools trained on the specific requirements of a government contract can monitor whether program delivery is tracking against contractual performance standards on an ongoing basis — alerting program managers to gaps before the reporting period ends, rather than discovering them during the report preparation process.

This shift — from reactive compliance documentation to proactive compliance monitoring — is the difference between a program that knows it met its targets and one that is discovering during report preparation whether it did.

Organizations that can monitor compliance continuously rather than reporting it quarterly are in a better position to course-correct when programs are underperforming. They are also in a better position to renegotiate contract terms, demonstrate capacity for expanded scopes of work, and defend their performance against audits.

Program Design and the Evidence Problem

There is a question that sits behind every program design decision, and it is rarely asked with enough rigor: What is the evidence that this intervention works?

The evidence base for nonprofit programming is uneven, to put it generously. Some program models — specific curricula, treatment protocols, intervention frameworks — have been rigorously evaluated and show consistent results across populations and settings.

Most have not. Most nonprofit programs are designed from a combination of practitioner experience, funder priorities, available funding streams, and organizational history. The connection to evidence is often asserted rather than demonstrated.

This is not a moral failing. Building an evidence base is expensive, methodologically demanding, and — in the real world of nonprofit funding — often less rewarded than the appearance of innovation. Funders will pay for new programs. They are less reliably interested in paying for rigorous evaluation of existing ones. The incentive structure has not historically favored evidence.

AI changes the calculus in two ways. First, it dramatically lowers the cost of identifying and synthesizing the existing evidence base for any given program area. What used to require weeks of literature review — surveying randomized controlled trials, systematic reviews, evaluation reports, and practitioner literature for a specific intervention — can now be accomplished in hours with AI-assisted research tools.

Program staff designing or refining programs have better access to the evidence than they have ever had before. The constraint is no longer the availability of the research. It is the organizational will to consult it.

Second, AI enables nonprofits to build their own evidence base more efficiently than was previously possible. Organizations tracking client outcomes with adequate data infrastructure can use AI to identify which program elements, delivered in which sequences, to clients with which characteristics, are most consistently associated with the outcomes they are trying to produce.

This is not a substitute for a randomized controlled trial. But it is far more informative than the alternative — which, in most nonprofits, is accumulated practitioner intuition unconnected to systematic data.

The funders who are moving toward evidence-based grant-making — the large national foundations, the federal demonstration grant programs, the impact investors — are increasingly asking organizations to demonstrate that their programs work and not merely that their staff believe in them.

AI-assisted program evaluation is not the answer to that challenge. Rigorous evidence is. But AI can help organizations develop and present that evidence more effectively than they have been able to do before.

Where Program Staff Should Start Today

The distance between where most program staff are today and where they could be with AI is significant. Closing it does not require a technology transformation initiative. It requires a series of deliberate, bounded first steps that build capability without demanding capabilities that don't yet exist.

The starting point is the reporting burden. Pick the most time-consuming grant report or government contract performance submission in the organization's portfolio. Map the data required to produce it. Identify where that data currently lives.

Then ask a simple question: Is the data that belongs in this report already captured somewhere in our systems? In most organizations, the answer is yes — it is simply not assembled automatically.

AI tools can assemble it. The first project is to make that happen for one report, complete it in half the time, and let the result make the case for expanding the approach.

The second starting point is outcome tracking. If program data is currently being captured in narrative case notes rather than structured data fields, AI tools can help extract structured information from those narratives at scale.

This is not a replacement for better data collection practices — it is a bridge to them. But it immediately creates a richer analytical foundation than the unstructured text alone provides.

The third starting point is early identification. Work with program data staff — or data vendors, or technology partners — to identify the three or four behavioral indicators in existing client data that are most predictive of disengagement or poor outcomes. Build a simple monitoring system around those indicators. Route weekly flags to case managers. Measure whether proactive outreach based on those flags changes outcomes for the clients who receive it.

This does not require a sophisticated machine learning model to begin. It requires data, discipline, and a willingness to act on what the data reveals.

Where Program Staff Could Go

The near-term applications are meaningful. The longer-term possibilities are transformative.

In five years, the organizations that have built real program data infrastructure — structured outcome tracking, integrated case management systems, consistent data collection practices — will be able to deploy AI in ways that organizations without that infrastructure cannot match.

Predictive models that identify clients likely to achieve specific outcomes with specific service packages. Real-time monitoring of program fidelity that ensures the intervention is being delivered the way the evidence suggests it should be. Adaptive programs that adjust in response to participant data without requiring a complete redesign cycle.

In the same time frame, the funders who support these organizations will increasingly expect this level of evidence. The national foundations that have been investing in data and evaluation capacity-building grants over the past decade are not doing so as a courtesy.

They are building toward a future in which the organizations that receive their largest grants are the ones that can demonstrate, with real data, that their programs work. AI-assisted evaluation will not be a competitive advantage for much longer. It will be the table stakes.

The program staff who develop AI fluency today — who understand what AI can and cannot do, who build the data practices that make AI useful, who learn to review and critique AI-generated reports and analyses rather than producing them from scratch — are building skills that will be increasingly indispensable.

Not because AI will replace program professionals. It will not. The judgment, the relationships, the ethical complexity, the cultural competence, the crisis response — none of that is algorithmic.

But the program professional who cannot work effectively with AI tools will be at a disadvantage relative to the one who can, in the same way that the fundraiser who refused to adopt email in the 1990s was at a disadvantage relative to the one who embraced it.

Mission Delivery Is the Mission

There is a version of the AI-in-nonprofits story that ends with a fundraising operation that is dramatically more efficient and a program operation that is essentially unchanged. That version is incomplete. It is also, ultimately, a failure of imagination.

The argument of this book is that AI in service of mission is powerful. If it is deployed only in service of fundraising, the sector has missed the larger opportunity. Programs are not the means to the fundraising end. Fundraising is the means to the program end. The people whose lives are changed by what nonprofits do are not a story to tell donors. They are the reason the organization exists.

AI can help identify them before they reach crisis. It can personalize the services they receive. It can measure whether those services are working. It can free the staff who serve them from administrative burdens that have been crowding out the actual work for decades. It can make the case to funders — with real data, in well-constructed reports — that the investment is producing what it promised.

Program staff who embrace this moment will find something unexpected: they will be better at their jobs. Not because AI is doing their jobs, but because it is doing the parts of their jobs that were never the point. The case manager who is no longer spending Thursday afternoon assembling a grant report has Thursday afternoon back.

What they do with it — the client they visit, the community partner they meet, the staff member they develop — is irreplaceable by any algorithm that has ever been built or ever will be.

That is the both/and. That is the opportunity. Not AI instead of mission. AI in service of it.

The tools will change. The mission won't.

16 FINANCE

"If you can't measure it, you can't improve it. But if that's all you do, you've missed the point." — Katharine Graham

Finance Folklore

There is a running joke inside most nonprofit organizations, and like most jokes that have survived for decades, it contains more truth than anyone is entirely comfortable admitting. *The finance department is where fun goes to die.*

Ask a fundraiser about their relationship with the finance office and you will hear some version of the same story. The budget requests that disappeared into a process nobody fully understood. The grant revenue that got coded wrong and triggered a two-week reconciliation. The end-of-year gift that arrived on December 31st and generated more paperwork than a capital campaign. The audit preparation season that consumed staff bandwidth for months and produced a document that sat in a binder on a shelf.

Ask a program director and you will hear different details but the same shape. The restricted fund that couldn't be spent on what actually needed to get done. The budget modification that required three approvals before anything could move. The funder financial report that required a format nobody in finance had seen before but that somehow still had to be submitted in nine days.

This is the accepted folklore of the sector: development raises the money, programs spend it, and finance stands in the middle making sure nobody does either one incorrectly. The relationship is necessary. It is not always warm.

Here is what the folklore misses. The finance department at a nonprofit is sitting on something valuable — something that development and programs desperately need and rarely have full access to: a complete, integrated, audited picture of the organization's financial reality.

Not the storytelling version. Not the version that gets presented to donors at a gala. The actual numbers. Where the money comes from. Where it goes. What it costs to do what the organization does. Which programs are financially sustainable and which ones are quietly subsidized by unrestricted reserves that the board doesn't realize are eroding.

That information is irreplaceable. It is also, in many organizations, trapped.

Trapped in accounting systems that nobody outside finance knows how to query. Trapped in spreadsheet models that a single staff member built over years and that nobody else fully understands. Trapped in audit reports that describe the past without illuminating the path forward. Trapped, ultimately, in the professional distance that has grown up between the people who track the numbers and the people who need to understand them.

AI doesn't eliminate that distance automatically. But it begins, for the first time in a serious way, to make the crossing easier — for the finance professionals who have always had the numbers and for the colleagues who have always needed them.

The Nonprofit Finance Reality

Most people outside the sector do not fully appreciate how structurally complex nonprofit finance actually is.

A community foundation managing donor-advised funds. A human services organization navigating forty government contracts, each with its own allowable cost schedule and draw-down timeline. A healthcare system with a foundation that must keep its philanthropic activity firewall-clean from the hospital's operational finances.

A university development office managing gifts that span permanently restricted endowments, temporarily restricted project funds, and unrestricted operating support — all in the same fiscal year, all subject to different rules.

This complexity predates AI by decades. It is the accumulated result of how the nonprofit sector was built: incrementally, often opportunistically, in response to funding available rather than organizational design preferred. Organizations didn't choose to have forty-three funds with forty-three different sets of restrictions. They accumulated them over twenty years of grant-seeking.

And now someone — usually a small finance team that hasn't grown proportionally with the organizational complexity it manages — has to hold it all together.

The people who do this work are not appreciated as much as they should be. They close the books every month. They manage the auditors every year. They track every dollar that flows through the organization against the restrictions it arrived with. They manage cash flow timing in organizations where revenue is episodic and expenses are continuous. They produce the financial statements that the board reviews for ten minutes in a meeting and approves with a motion.

They do all of this — and they do it, in most nonprofits, with financial systems that were not designed for the specific complexity of the nonprofit environment, with staff sizes that have not kept pace with organizational growth, and with technology that, until very recently, had not materially advanced the underlying work.

The spreadsheet that finance uses to track grant drawdowns today is, in most organizations, structurally identical to the one they used fifteen years ago.

The names and numbers have changed. The method has not. AI changes the method.

What the Numbers Actually Know

Before exploring where AI applies, it is worth pausing on what the finance function actually produces — and what that production makes possible.

The general ledger is not just an accounting record. It is an organizational memory. Every restricted gift. Every program expense. Every indirect cost allocation. Every year-end reclassification. Every audit adjustment. Encoded in the ledger, over years, is a detailed history of what the organization did, what it cost, and where the resources came from. This history is more complete, more precise, and more auditable than any other record the organization maintains. It is also, in most nonprofits, effectively inaccessible to the people who most need it.

The development director who wants to understand the true cost of a major gift campaign — not just the direct expenses but the fully-loaded cost including staff time, benefits, overhead allocation, and technology — cannot easily get that number from the finance system.

The program director who wants to compare the cost per client served across three different service delivery models cannot easily get that number either. The CEO who wants to understand which revenue streams are growing and which ones are quietly declining faces a version of the same problem: the data exists, but the path from data to insight requires either deep familiarity with the finance system or a conversation with the finance director that has to be scheduled two weeks in advance.

This is not a character flaw. It is a system design problem. Finance systems are designed to produce accurate records, not accessible insights. They are optimized for audit defensibility, not for the kind of analytical querying that strategy requires.

And the finance professionals who operate them are, in most nonprofits, fully occupied managing the systems themselves — leaving little bandwidth to translate what those systems know into guidance for the colleagues who could benefit from it.

AI shifts that equation in both directions: it makes the finance function itself more efficient, which creates bandwidth; and it makes financial information more accessible to the people who need it, which creates value. The two effects compound.

What AI Can Do for the Finance Function Today

The immediate applications are not speculative. They are available now, adopted by nonprofit finance teams that have decided to invest the time to implement them.

The first application is budget development and forecasting. Building a nonprofit budget is not the mechanical exercise it appears to be from the outside. It requires synthesizing assumptions about donor behavior, grant renewal probabilities, program growth trajectories, staffing plans, and indirect cost allocations — and producing a document that is defensible to leadership, to the board, and, in some cases, to funders who have approved budgets as part of grant agreements.

A finance director building next year's budget is making dozens of judgment calls, often working from historical data that is partially available in the accounting system and partially held in the heads of the people who made the decisions that produced it.

AI tools can dramatically improve this process. They can ingest historical financial data — revenue by stream, expenses by program, cost trends over multiple years — and produce baseline projections that serve as a starting point for human judgment rather than requiring that the starting point be built from scratch.

They can model scenarios: What does the budget look like if the federal contract doesn't renew? What happens to operating reserves if the gala raises 20% less than last year? What is the break-even point for a new program that requires hiring two full-time staff before it can generate revenue? These scenarios take hours to model manually.

With AI, the finance director can run them in minutes — not because AI has replaced the judgment required to evaluate the scenarios, but because it has eliminated the spreadsheet labor required to construct them.

The second application is grant financial management. A medium-sized nonprofit managing twenty active grants has twenty separate financial management obligations: tracking allowable and unallowable expenditures, managing budget-to-actual variances within each award, meeting drawdown deadlines, preparing financial reports in funder-required formats, and flagging when a grant is at risk of under- or over-spending before it becomes a compliance problem.

This is not intellectually complex work. It is administratively exhausting work — exactly the kind of high-volume, rules-based, deadline-driven management task where AI performs reliably and humans tire.

AI-assisted grant financial management systems can monitor spending against budget in real time, flag variances that require attention, generate draft financial reports in funder-required formats, and alert staff when a grant is approaching critical thresholds.

The finance staff member who was spending two days every quarter generating grant financial reports can spend that time on the analysis and funder relationships that matter more. The program staff who were anxiously tracking their own mental models of where each grant stood can see the picture clearly and act on it before problems develop.

The third application is accounts payable and expense management. Nonprofit finance teams spend meaningful amounts of staff time on the mechanics of payment processing: reviewing invoices, coding expenses to the right fund and program, obtaining approvals, cutting checks or initiating wires, and reconciling accounts.

AI tools with document processing capabilities can now read invoices, extract key data, propose proper coding based on historical patterns, and route transactions for approval — with human review focused on the exceptions and edge cases rather than the routine processing.

The finance team that once processed two hundred invoices per month with a four-day turnaround can process the same volume in half the time, with greater coding accuracy and fewer reconciliation exceptions.

The fourth application is financial reporting and board communication. The monthly financial package that goes to the board — the one that finance staff spend two days assembling from data that could theoretically be pulled in twenty minutes — is overdue for reinvention.

AI tools can generate board-ready financial summaries directly from accounting data, formatted consistently, with variance analysis against budget included, and narrative commentary drafted from the underlying numbers. The finance director reviews, adds context where judgment is required, and presents — rather than preparing and presenting.

More importantly, the financial narrative that actually informs board decision-making — not just the compliance-oriented statement of activity, but the real picture of organizational financial health — can be produced with greater depth and clarity when finance staff are no longer consumed by assembly.

The board that currently reads a six-page financial report with limited context can receive a more insightful analysis of where the organization stands, where it is going, and what decisions require attention. That is not a cosmetic improvement. It is a governance upgrade.

The Bridge Function

Here is the application that nonprofit finance leaders have not yet fully claimed — and that may matter more than any efficiency gain.

Finance has always been, in theory, the department that connects the mission to the money. What are we trying to accomplish? What does it cost? Do we have the resources to sustain it? Are we deploying those resources in ways that are consistent with our strategy and our obligations?

These are finance questions. They require financial answers. And they are, in most nonprofits, answered incompletely — not because the finance function is incapable, but because the demands of compliance and operations leave insufficient time for the strategic work.

AI can close that gap.

Consider the development office. Fundraisers think about revenue. Finance thinks about the full picture of revenue and cost. A major gift that costs $50,000 in staff time and cultivation expenses to close — against a $75,000 gift — looks very different from a major gift that closes at the same amount with half the investment.

Development offices rarely track this. Finance, in theory, has the data to calculate it. In practice, the calculation doesn't happen because nobody built the connection between the fundraising activity data and the financial records that would make it possible.

AI-assisted analytics can now build that connection. Cost of fundraising by channel. Return on investment by program and donor segment. The true cost per dollar raised for each component of the development operation — annual fund, mid-level, major gifts, planned giving, events. These numbers are not just interesting. They are strategic instruments.

The development director who knows that her major gift program returns $12 for every dollar invested while the gala returns $2.80 has a powerful argument for how to allocate staff time and expense in the next budget cycle.

The board that sees both numbers understands something about organizational strategy that they have never been shown before.

The same connection exists between finance and programs. Cost per client served. Cost per outcome achieved — not just the number of clients who went through the program, but the measurable change in their circumstances. Program margins — which activities generate a surplus of grant and earned revenue over direct cost, and which ones are cross-subsidized.

These are not abstract metrics. They are the financial architecture of the mission. Organizations that track them can make strategic decisions about program mix, capacity, and investment. Organizations that don't are flying without instruments.

The finance department that deploys AI to build these connections — that moves from the compliance and reporting function it has always performed toward the analytical bridge function it has always had the potential to perform — becomes something the sector desperately needs more of: a strategic partner.

What Finance Leaders Need to Do Differently

The technology is not the barrier. The barrier is the role that finance has accepted in most nonprofit organizations — and the decision that finance leaders need to make about whether that role is the right one going forward.

Finance has accepted, largely without objection, the position of compliance custodian. The people who ensure that the money is accounted for correctly. The people who say no to things that would cause audit problems. The people who explain to the board that the organization is not technically insolvent and that the restricted funds cannot be accessed for operating shortfalls.

These are important functions. They are also insufficient ones, given what nonprofit organizations are actually up against.

The finance department that uses AI to become more efficient at its current role — processing faster, reporting more accurately, reducing reconciliation time — has captured the smaller prize. The finance department that uses the efficiency AI creates to build the analytical and strategic functions that the sector has been missing will have changed its organizational role entirely.

The first version is better. The second version is transformative. That transformation requires three things.

First, it requires finance leaders who understand that their job is not just to track what happened — but to illuminate what is happening and what is likely to happen next. Predictive financial modeling. Scenario planning. Sensitivity analysis on key revenue assumptions.

These are not exotic capabilities. They are the applications of existing financial knowledge to forward-looking questions. AI makes them tractable. The finance leader has to decide they are part of the job.

Second, it requires a deliberate decision to improve the financial literacy of the whole organization. Development professionals who don't understand the difference between cash and accrual accounting make decisions that cause reconciliation problems.

Program directors who don't understand restricted versus unrestricted funds make commitments that create compliance risk. Board members who nod along to financial statements they don't fully understand cannot govern effectively.

Finance can change this — not by making everyone an accountant, but by translating financial reality into language that the rest of the organization can act on.

AI-assisted communication tools make this easier. The decision to prioritize it is a human one.

Third, it requires building the data infrastructure that makes AI analysis possible. The organization that tracks grant expenditures in spreadsheets, maintains its donor financial history in a separate system from its accounting records, and reconciles its general ledger to its grant management system manually is not ready to deploy the advanced analytics that AI makes possible.

Fixing the infrastructure is a precondition for realizing the benefit — not a someday project, but a prerequisite.

The Treasurer's Seat at the Table

There is a version of the future where the nonprofit finance director is one of the most strategic people in the organization. Where their insights inform development strategy, program investment decisions, board governance, and organizational planning.

Where the financial picture they maintain is not a backward-looking compliance record but a live model of organizational health that leadership uses to make decisions in real time.

That version is closer than it appears. The technology to build it exists today. The financial expertise to use it already lives in the finance department. What remains is the organizational decision to invest in building the bridge — to connect the financial intelligence that finance has always held to the strategic conversations it has too rarely joined.

The finance department that claims this role will be a different kind of organizational partner than the one that has been running the same playbook for the last twenty years.

It will ask harder questions. It will surface uncomfortable truths. It will tell the development team that their most beloved fundraising event is not returning what they think it is. It will tell program leadership that a cherished initiative is costing twice what comparable programs cost elsewhere and producing half the measurable outcomes.

These are not easy conversations. They are necessary ones. And the organizations that have them — regularly, honestly, with real data — will make better decisions than the organizations that don't.

The foundation was always there. The numbers always knew.

AI is giving the sector the tools to finally hear what the numbers have been trying to say.

The finance director who embraces that role is not the person who says no to things. She is the person who says: here is what the data shows, here is what it means, and here is what we should consider doing about it.

That is not the finance function the sector is used to. It is the one the sector needs.

The numbers have always been there. Now, for the first time, there is no good reason not to listen to them.

17 REVENUE

"Revenue is the oxygen of the organization. Without it, every other plan suffocates." — Frances Hesselbein

Nonprofit Does Not Mean No Profit

There is a persistent myth that haunts the nonprofit sector. It is rarely spoken aloud — that would be too direct — but it shapes budgets, constrains ambitions, and quietly undermines the financial health of thousands of organizations every year.

The myth goes like this: nonprofits are different from businesses. They exist to do good, not to make money. And the pursuit of financial strength is, at best, a necessary distraction from the real work.

This is wrong. It is worth saying plainly: Being a nonprofit does not mean there is no profit.

It means that any surplus is reinvested in the organization for the benefit of the mission — not distributed to shareholders. In the United States, these are tax-exempt organizations under Section 501(c)(3) of the Internal Revenue Code.

That designation describes a tax status. It does not describe a financial philosophy, and it certainly does not exempt an organization from the requirement to generate revenue in excess of expenses.

An organization that perpetually breaks even — that ends every fiscal year with nothing left over — has no reserves to weather disruption, no capacity to seize opportunity, and no cushion against the inevitable surprises that every organization eventually faces. Perpetual breakeven is not mission purity. It is financial fragility dressed up as virtue.

The sector has an old phrase for this reality. No money, no mission. Or, if you prefer the more direct formulation: No profit, no nonprofit.

Both versions are saying the same thing, and the most effective nonprofit leaders have internalized it. They do not treat financial performance as a regrettable concession to the practical world. They treat it as a moral obligation — because an organization that cannot sustain itself financially cannot keep its commitments to the communities it serves.

Understanding revenue in a nonprofit requires understanding its full structure, which is more complex than most people outside — and many people inside — the sector appreciate.

More than 80 percent of all nonprofit revenue in the United States comes from fee-for-service income and government grants and contracts. Charitable contributions, which occupy the majority of the development office's attention, represent roughly 14 percent of total sector revenue. Investment income accounts for approximately 5 percent more.

The general public tends to focus on donations and grants, but the bulk of what keeps nonprofit organizations running comes from somewhere else entirely.

This matters because AI's opportunity in the revenue domain is not confined to the development office. It runs through the entire organization.

Revenue in a nonprofit works across four fundamental dimensions. Getting it. Keeping it. Growing it. And netting it — because gross revenue that disappears into unnecessary expense does not fund a mission.

Each dimension has its own disciplines, its own failure modes, and its own AI applications. Leaders who understand all four can build an organization that is not just impactful today but financially sustainable across decades.

Get Revenue

Every organization, no matter how established, must continuously acquire new revenue. Donors lapse. Grants end. Government contracts are terminated or restructured. Fee-for-service programs lose participants. The revenue base erodes constantly, and acquisition is the work of filling the bucket faster than it leaks.

The acquisition challenge looks different depending on where the revenue comes from.

For charitable contributions, acquisition begins with finding people who have both the capacity to give and a genuine connection to the mission — and who have not yet been asked.

Most organizations are sitting on this population without knowing it. The constituent database is full of people who have attended events, volunteered, subscribed to communications, toured facilities, or benefited from programs who have never been converted to donors.

The traditional approach to donor acquisition — buying mailing lists, running broad digital campaigns, sponsoring peer-to-peer fundraisers — treats this warm population the same way it treats complete strangers. That is an expensive mistake.

AI changes the front end of acquisition by making it far more precise. A model trained on the behavioral and demographic profile of current donors can scan the unconverted constituent population and identify probable look-alikes with a granularity that no manual review process could approach.

The event attendee who has been coming to the annual gala for four years, opened every email, and referred two colleagues to the volunteer program looks nothing like a cold prospect. AI can tell you that.

The major gift officer who receives a prioritized list of high-affinity unconverted constituents — each one flagged with the specific signals that suggest readiness for a first ask — walks into a conversation that is already warmer than any cold outreach could produce.

For earned income, acquisition means expanding the number of participants, patients, students, or clients the organization serves — and doing so without proportionally increasing cost.

A nonprofit vocational training organization that certifies workers for skilled trades is bounded by how many participants it can enroll and how many instructors it can deploy.

AI-assisted instruction can expand the instructor-to-participant ratio significantly. More certifications issued. More tuition revenue collected.

The same mission delivered to a larger population with the same or lower unit cost. This is not an abstraction. It is a revenue growth strategy with a direct line to the income statement.

For government grants and contracts, acquisition means winning new awards in a competitive market. Grant writing for government funders is among the most demanding forms of institutional communication — proposals that run hundreds of pages, compliance requirements specified in exhaustive detail, evaluation criteria that reward the precise alignment of organizational language with funder priorities.

Organizations that lack dedicated grant development capacity lose opportunities to better-resourced competitors, not because their programs are weaker but because their proposals are.

AI can close a meaningful portion of that gap — accelerating the production of narrative sections, synthesizing outcome data into the format the solicitation requires, reviewing draft proposals for responsiveness to the evaluation criteria, and flagging gaps before the deadline rather than after the rejection.

The organization that can respond to twice as many solicitations with the same staff has doubled its acquisition surface.

The unifying principle across all three channels is this: AI does not create revenue from nothing. It amplifies the capacity of organizations that already have something worth funding to find and secure more of the support they deserve.

Keep Revenue

Getting revenue once is an accomplishment. Getting it again is the business model.

Donor retention — the percentage of contributors who give a second time, and a third, and a tenth — is one of the most powerful levers in fundraising, and one of the most systematically undervalued.

The math is unforgiving. First-year donor retention for the average nonprofit sits at approximately 29%. Seven out of ten new donors never give a second time.

For every dollar invested in acquiring that lapsed donor, the organization received a single gift and then spent more money trying to win them back — usually at a single-digit reactivation rate.

The acquisition expense is visible on the budget. The compounded lifetime value of the relationships that were never built is invisible, which makes it easy to ignore and ruinous to accept.

A donor retained for ten years is not ten times more valuable than a donor retained for one. Because retained donors upgrade, introduce others, respond to major gift cultivation, and ultimately become planned giving candidates, the multiplier is far larger.

The first gift is the beginning. Every subsequent gift is a return on the relationship investment made between asks.

AI keeps revenue primarily by keeping relationships from going cold before anyone notices.

Lapse prediction is the most immediately valuable application. A model trained on the behavioral patterns of donors who lapsed in previous years — declining response rates, decreasing gift amounts, increasing time between gifts, disengagement from non-giving touchpoints — can flag at-risk relationships months before the actual lapse.

That window is everything. A personal call from a staff member to a donor who has given for eight consecutive years, made before the lapse rather than after, carries a fundamentally different emotional register than a reactivation appeal sent to someone who has already been gone for two years.

One is relationship maintenance. The other is damage control. The costs are different. The success rates are very different.

Retention of government contract revenue requires a different discipline entirely: compliance. Federal contracts in particular impose detailed performance standards, quarterly and annual reporting requirements, and financial documentation obligations that consume enormous staff time — not because every compliance requirement is meaningful, but because the consequences of non-compliance are severe enough that organizations rationally over-invest in documentation to protect themselves.

Program staff become part-time compliance officers. Bandwidth that should flow toward service delivery flows toward paperwork instead.

AI can absorb a meaningful share of this burden. Compliance reporting workflows that currently require manual data extraction, reconciliation, and narrative drafting can be automated to the point where a first draft of a quarterly performance report requires review and approval, not creation from scratch.

More importantly, AI tools trained on the specific requirements of a government contract can monitor ongoing program performance in real time — alerting program managers to gaps before the reporting period ends, rather than discovering them during report preparation. This shift from reactive compliance documentation to proactive compliance monitoring is the difference between a program that knows it met its targets and one that is hoping it did.

For fee-for-service revenue, retention means keeping participants enrolled, keeping clients engaged, and keeping payers — whether individuals, government agencies, or insurers — satisfied with the quality of what they're purchasing. AI-assisted program delivery improves quality without proportionally increasing cost, which is the most fundamental retention argument available: people stay when the service is worth staying for.

Grow Revenue

Getting revenue and keeping revenue are the essential functions. Growing revenue is the strategic ambition — and the difference between an organization that sustains its current capacity and one that expands it.

Growth in charitable giving follows a well-documented pathway. Annual donors who are properly cultivated become mid-level donors. Mid-level donors who are properly identified become major gift prospects. Major gift donors who are deeply stewarded become planned giving candidates.

The pathway is known. What most organizations lack is the systematic capacity to see who is on it and to act at the right moment before the opportunity passes — before a donor ages into estate giving without anyone recognizing the signal, before a mid-level donor's circumstances change in ways that suggest major gift capacity, before a first-year donor whose profile matches a future loyalist drifts away because nobody followed up.

AI keeps the full continuum visible in a way that no manual process can match at scale. The loyal annual fund donor who has given for twenty-three consecutive years, whose demographic profile and absence of major gift activity despite apparent capacity suggests estate planning potential, can be identified before the bequest arrives out of nowhere — which is to say, before it almost doesn't arrive at all.

The first-year donor whose behavioral signals match the documented profile of long-term loyalists gets flagged while the relationship is young enough to shape deliberately. The mid-level donor who attended four events last year, opened every impact update, and whose wealth screening suggests capacity far above their current giving level gets surfaced before a peer institution's gift officer gets there first.

Planned giving deserves particular emphasis in any conversation about revenue growth, because the gap between potential and practice is so large.

Approximately half of all endowment growth in the nonprofit sector traces back to donors who chose to include an organization in their estate plan. The donor who leaves a $2 million bequest is, in most cases, a $100-a-year annual fund contributor whose generosity was never suspected and whose estate plans were never discussed.

That donor is not rare. They are common — and they are invisible to every organization that relies on wealth screening and demographic guesswork to identify planned giving prospects, because modest annual giving obscures enormous deferred capacity.

AI models trained on the behavioral and giving profiles of confirmed planned giving donors can identify likely candidates across a large file with a precision that transforms the planned giving program from a long-shot lottery into a systematic practice.

Revenue growth in earned income follows a different logic but arrives at a similar conclusion. Fee-for-service programs that use AI to expand capacity — serving more participants at lower unit cost — grow their top line without requiring proportional resource investment.

The margin on those additional participants is higher than the margin on the first participants, because the fixed costs are already covered.

That additional margin is unrestricted income — the most valuable kind — and it can flow wherever the mission requires it most, including toward the programs that cannot generate revenue at all.

The organizations that grow revenue consistently are not the ones that happened to hire the best fundraisers. They are the ones that built systems capable of seeing their own opportunities clearly enough to act on them before they close.

Net Revenue

Every conversation about revenue eventually has to pass through the same doorway: what's left.

Gross revenue is the number that goes on the press release. Net revenue — what remains after expenses — is the number that determines what the organization can actually do. And here is where many nonprofits, even well-intentioned ones run by capable people, allow a cultural discomfort with financial management to produce genuinely poor results.

If gross revenue is discussed proudly and openly, expense control is often discussed reluctantly, defensively, or not at all.

The result is a sector where a remarkable number of organizations raise more money every year than they raised the year before and still end the year in a weaker financial position.

Expenses grow to match or exceed revenue. Reserves erode. The operating cushion that would allow strategic investment or risk-taking disappears. The cycle repeats.

Net revenue requires two things: a clear understanding of what things actually cost, and the discipline to manage those costs against a strategic framework. Both are harder than they sound in organizations where finance is treated as a back-office function rather than a strategic partner — where the general ledger holds an organization's complete financial memory but that memory is effectively inaccessible to the people who most need it.

AI is changing that equation in ways that nonprofit finance leaders are only beginning to claim.

The cost of fundraising is a number most development operations do not know with precision, and the lack of precision is doing real strategic harm.

Mature development operations typically invest 15 to 25 percent of funds raised in advancement costs — salaries, technology, communications, events, and cultivation expenses. But the total number is not the strategically useful number.

What matters is the cost by channel, by program, and by donor segment. The major gift program that returns twelve dollars for every dollar invested looks nothing like the gala that returns two dollars and eighty cents.

A development director who knows both numbers has a powerful argument for how to allocate staff time and budget in the next planning cycle. A development director who knows only the aggregate is making resource allocation decisions without the data those decisions require.

AI-assisted financial analytics can now build the connection between fundraising activity data and financial records in ways that manual processes never could. Cost per donor acquired by channel. Return on investment by program component. The fully-loaded cost of a major gift — including not just direct expenses but the staff time, travel, cultivation events, and stewardship communications that preceded the close — compared against the gift amount.

These calculations are not interesting as academic exercises. They are operational instruments that separate organizations managing their fundraising investment strategically from those managing it by intuition and habit.

The same principle extends to programs. Cost per client served. Cost per outcome achieved — not just the number of people who passed through the program, but the measurable change in their circumstances. Which programs generate a surplus of grant and earned revenue over direct cost, and which ones are quietly cross-subsidized by unrestricted reserves the board doesn't fully realize are eroding.

These are the financial architecture of the mission, and organizations that can see that architecture clearly can make decisions about program mix, capacity, and investment that organizations operating without it cannot.

Budget forecasting — the work of projecting what revenue will be available and what expenses must be managed against it — is the discipline that connects the other three revenue dimensions to net revenue.

AI tools can ingest historical financial data across multiple years, model scenarios with a speed that was previously impossible, and produce the kind of multi-variable analysis that used to require either a sophisticated finance team or expensive external consultants.

What happens to reserves if the federal contract doesn't renew? What is the break-even point for a new program that requires two full-time hires before it generates revenue? What does the cash position look like if the major giving campaign runs six weeks behind target? These scenarios take hours to model manually.

With AI, the finance director runs them in minutes — not because AI has replaced the judgment required to evaluate the scenarios, but because it has eliminated the spreadsheet labor required to construct them.

Organizations that invest in this financial clarity will make better decisions at every level. The development director who understands true cost of fundraising by channel will invest more wisely. The program director who can see cost per outcome will design more efficiently. The CEO and board who can see the full revenue picture — not just the gross numbers, but the net position and the trajectory — will govern with a precision that organizations flying without instruments cannot approach.

Net revenue is not an accounting formality. It is the score. And leaders who understand the score — who refuse to confuse activity with results, or gross revenue with financial health — are the ones building organizations capable of sustaining their missions across generations.

The Full Picture

Get it. Keep it. Grow it. Net it.

These are not four separate strategies. They are four lenses on the same fundamental challenge: building an organization that is financially strong enough to do what it says it will do, for the people who are counting on it, for as long as the mission requires.

The cultural discomfort with financial ambition in the nonprofit sector is understandable. Organizations founded to serve others are right to be suspicious of the drift toward self-preservation for its own sake.

But financial strength in service of mission is not self-preservation. It is mission fulfillment at scale. The organization that ends every year with no surplus, no reserves, and no cushion is not being virtuous. It is being one bad year away from failing the people it exists to serve.

AI does not make the hard strategic choices for you. It does not decide which programs to fund or which relationships to prioritize.

It does not replace the leadership judgment that turns financial data into organizational direction. What it does — consistently, scalably, and with a precision that no previous generation of leaders had access to — is illuminate the picture clearly enough that the choices become visible.

No money, no mission. The tools have changed. The truth hasn't.

18 TRUST

"Trust is the lubrication that makes it possible for organizations to work." — Warren Bennis

Change Requires Changing

There is a moment in the adoption of every genuinely new technology when the argument stops being about capability and starts being about something far more fundamental. The engineers have proven it works. The early adopters have demonstrated it delivers. The business case has been made, the return on investment calculated, the implementation plan drafted.

And still — the answer is no. Not yet. Not us. Not this.

The question is this: Can I trust it?

Not — does it work? The demonstrations have been convincing. Not — is it affordable? The pricing has come down.

Not — has anyone else done this? The case studies exist. The question that stops the implementation plan is older and harder than any of those, and it has almost nothing to do with software.

Can I trust it?

That question is not answered by a vendor's data sheet. It is not answered by a reference call with a peer organization. It is answered — slowly, experientially, and only one way — by the accumulated weight of evidence that the thing being trusted has earned the trust being extended.

That process cannot be accelerated by enthusiasm. It cannot be bypassed by authority. It cannot be mandated from a board resolution or a CEO announcement.

Trust is not installed. It is built. And the people who don't understand that are the ones who keep wondering why their AI systems have a 30 percent adoption rate.

People like change. They do not like changing. This is not a contradiction — it is one of the most consistent findings in the study of human behavior. We celebrate innovation in the abstract. We resist it in the specific. We admire organizations that transformed themselves. We avoid transforming ours. The dissonance is not hypocrisy. It is biology.

Understanding why trust is the central challenge in AI adoption requires understanding something deeper than technology strategy. It requires understanding what the human brain is actually built to do — and what it was never designed to handle.

The Oldest Operating System

The human brain reached something close to its current form approximately 300,000 years ago in the middle Pleistocene. For roughly 295,000 of the years since then, the primary cognitive challenge facing any individual human was not strategic planning, organizational design, or digital transformation. It was survival.

Specifically, it was the problem of correctly distinguishing — in conditions of extreme time pressure, with significant information asymmetry and severe consequences for error — between what was safe and what was dangerous.

The neural architecture that evolved to solve that problem is still running inside every person sitting in your next all-staff meeting.

The amygdala — the almond-shaped structure buried in the brain's temporal lobe — is the ancient threat detection system. It processes incoming sensory information approximately 50 milliseconds faster than the prefrontal cortex, which is the region responsible for the kind of rational analysis that produces sentences like "AI will increase our fundraising efficiency by 23 percent."

This means that before any conscious evaluation begins, the brain has already registered a threat assessment. Before the argument for the new tool has been heard, a judgment has been rendered.

The instinct fires first. The logic follows — and often, its real function is not to evaluate the original question but to justify the conclusion the amygdala already reached.

This is not a cognitive failure. It is a design feature — one that was exquisitely adapted to an environment where the consequences of extending trust incorrectly were measured in blood rather than budget cycles.

The creature that approached the unknown with curiosity and openness was sometimes rewarded. It was also sometimes eaten. The creature that approached the unknown with suspicion and caution missed some opportunities. It also survived long enough to have children.

Evolution is a rigorous editor, and over 300,000 years it edited the human brain to treat novelty as a threat until proven otherwise.

The problem — the one every technology deployment in human history has had to solve — is that this ancient threat-detection software is running on hardware that now occupies a world of artificial intelligence, predictive analytics, and machine learning tools.

The amygdala that evolved to assess tigers and unfamiliar strangers is now assessing a CRM platform's AI-powered prospect scoring feature. It cannot tell the difference.

Novel, opaque, and consequential: those were the characteristics of the predator in the grass, and they are also, from the brain's perspective, the characteristics of an algorithmic system it doesn't understand.

The resistance to AI in organizations is not a technology problem. It is a neuroscience problem. And solving it requires building trust the way the brain is actually wired to receive it — not through argument, but through experience; not through assertion, but through evidence; not through one dramatic revelation, but through the accumulation of small, reliable, predictable confirmations that the thing being trusted is what it claims to be.

The Machine Breakers Were Not Wrong

In the winter of 1811, a group of skilled textile workers in Nottinghamshire began destroying the power looms and stocking frames that were eliminating their livelihoods. The movement spread rapidly across the industrial Midlands and the North of England.

By the following spring, the British Parliament had passed the Frame Breaking Act, making machine-breaking a capital offense — a penalty more severe than what was then prescribed for a great many murders.

The army deployed against the Luddites at the peak of the unrest numbered more soldiers than Wellington had committed to the Peninsular War in Portugal.

History has not been kind to the Luddites. Their name has become the shorthand for anyone who resists technology out of ignorance or fear, and the shorthand has stuck long enough that most people have forgotten what the Luddites were actually resisting.

They were not resisting technology they didn't understand. They were resisting, with entirely accurate foresight, the economic displacement of skilled workers by mechanized production. The hosiers and croppers who led the movement understood — correctly — that the power loom was going to make their expertise worthless, their wages uncompetitive, and their communities hollowed out.

They were right. The question they got wrong was not the diagnosis. It was the prescription. Destroying the machines did not stop the transformation. It only made the people doing the destroying into criminals.

The lesson that subsequent technological revolutions failed to learn — the one worth learning now — is that the people who resist are often seeing something real. The fear embedded in resistance is not always irrational. The skilled textile worker who understood that the loom was going to make thirty years of expertise irrelevant had perceived a genuine threat.

The nonprofit fundraiser who looks at an AI-powered donor scoring system and wonders whether it will make her relationships redundant is not being paranoid. She is conducting a threat assessment. The amygdala is doing its job.

The organizations that drove through that resistance without engaging it did not build trust. They built compliance and resentment — two things that look like adoption in the short run and collapse under stress.

The organizations that addressed the genuine concerns embedded in resistance — that took the threat assessment seriously, examined it honestly, and either addressed it or explained why it was mistaken — those organizations built something more durable.

They built the kind of trust that survives the first failure, because the people in the organization understood the system well enough to distinguish between a failure mode worth worrying about and a normal error rate in a system that was still doing its job.

Edison, Ford, and the Mechanics of Trust

Thomas Edison understood the trust problem before he understood much about commercial electricity. He opened the Pearl Street Station in lower Manhattan in September of 1882 not because New York was the optimal location for a power generation facility — it wasn't — but because lower Manhattan contained the offices of the most influential newspapers in the country.

Edison wanted J.P. Morgan, newspaper editors, and New York's most visible business leaders to experience electric light firsthand. He wanted witnesses. And he understood — with a commercial intuition that was as sophisticated as his technical intuition — that trust spreads through personal experience far faster than it spreads through argument.

The first customers of the Pearl Street Station were not consumers. They were a trust infrastructure. Edison had calculated that if the right people experienced electric light working reliably and safely, their testimony to everyone they knew would accomplish what no advertisement could. He was right.

But the mechanism only worked because the system actually performed as promised, consistently, in conditions where failure would have been publicly catastrophic. The moment a system fails visibly — especially early in its deployment, especially in front of the people whose confidence you most need — you don't just lose that installation. You lose the whole argument.

Edison's obsession with reliability was not perfectionism. It was a trust strategy. He understood that every successful demonstration was not just a commercial transaction. It was a deposit in the account of public confidence. And he understood, with equal clarity, that a single dramatic failure could withdraw everything he had deposited.

The asymmetry of trust — easy to lose, slow to earn, essentially impossible to manufacture — shaped every technical and commercial decision he made in the first decade of the electrical age.

Henry Ford understood the same principle and expressed it differently. The Model T was not the most powerful automobile of its era, or the most elegant, or the most technically sophisticated. It was the most reliable.

Ford's relentless focus on manufacturing precision — on tolerances so consistent that any component could be swapped into any car on the assembly line — was, at its deepest level, a trust strategy.

People who had no idea how an internal combustion engine worked trusted the Model T because it kept running. Every time it kept running, the trust deepened. Every time any automobile failed catastrophically and publicly, the trust in the entire category eroded a little.

Ford understood that trust in a new technology is never singular — it is categorical. When your AI-powered donor communication system sends a message with the wrong name to a longtime major donor, the donor does not lose trust in that specific system configuration. She loses trust in AI-assisted communication. Possibly in AI more broadly.

The failure propagates outward from the specific incident because the brain's threat-detection system generalizes from specific experiences to categorical conclusions. One bad experience with a novel technology teaches the brain that the category is dangerous. Multiple reliable positive experiences teach the brain that the category is safe. The math is not fair. It never has been.

The organizational implication is direct. Start small. Measure honestly. Fix publicly what breaks. Do not overstate what the system can do. Do not paper over early failures in the enthusiasm to demonstrate progress.

The organizations that will have trusted AI in five years are the ones that deployed it carefully in the first year — not because caution is inherently virtuous, but because every early success is a deposit and every overstated capability is a debt that will be collected at the worst possible moment.

The Opacity Problem

AI presents the trust challenge in a form that has no clean historical precedent, and nonprofit leaders deserve an honest account of why.

You can see how a loom works. You can trace the logic of a spreadsheet from input to output. You can open an internal combustion engine and understand, at least in principle, the sequence of events that produces motion.

The relationship between action and result in most previous technologies was visible, or at least traceable. When something went wrong, you could usually find out why.

The machine learning models that power modern AI applications do not work this way. The relationship between input and output passes through layers of mathematical abstraction that are opaque even to the engineers who built them.

When a prospect scoring model recommends a particular donor for immediate major gift cultivation, it can surface the features that weighted most heavily in the calculation. It cannot narrate its reasoning the way a human analyst can.

The architecture of the decision is not accessible in the way that mechanical or logical systems are accessible. The box is genuinely black.

This opacity creates a specific trust problem. The people who need to trust AI outputs in order to act on them — the gift officer deciding whether to prioritize a prospect, the program director evaluating a risk flag, the development director reviewing a lapsed donor list — are being asked to rely on recommendations they cannot fully trace.

The confidence of the model is expressed as a number. The reasoning behind the number is not something they can interrogate with the same directness they would apply to a human colleague's recommendation.

This is a real problem. It deserves an honest answer rather than reassurance, and the honest answer is this: opacity is not unique to AI. It is the common condition of every important institutional trust relationship.

You cannot trace every transaction your financial institution processes on your behalf. You cannot verify in real time every decision your organization's legal counsel makes. You extend trust to the physician based on their credentials, their track record, and your experience with their judgment — not based on your ability to follow the biochemistry of their diagnosis.

At some level, every consequential institutional relationship involves extending trust across an opacity gap. What matters is not whether the gap exists. It will always exist.

What matters is what is built on either side of it — what oversight mechanisms, what accountability structures, what performance tracking, what transparency about limitations — to give the people being asked to trust something solid enough to stand on.

The organizations that have built genuine AI trust are not the ones that made the black box transparent. They are the ones that built the scaffolding around it that makes the black box safe to rely on.

What Personalization Becomes Without Trust

Chapter 4 named the anxiety. This chapter is about the specific relationship dimension where that anxiety finds its sharpest expression: the intersection of AI capability and donor trust.

Donors give to people and causes they trust. This is not a sentiment. It is one of the best-documented findings in the study of philanthropic motivation, and it has been confirmed by every rigorous analysis of fundraising effectiveness since Ward and Pierce ran their first systematic campaigns more than a century ago.

The cause matters. The relationship matters. The conviction of the person making the ask matters. Remove any of those three elements and the gift is substantially less likely to happen — and substantially less likely to be renewed when the next solicitation arrives.

Now consider what a sophisticated AI deployment looks like from the donor's perspective. The communication she receives this week was drafted by an AI system, reviewed and lightly edited by a staff member she has never met, and sent under the signature of the development director she has met once.

The prospect score that determined she was ready for a personal visit was calculated by a model she doesn't know exists, using behavioral data she provided in contexts she has largely forgotten.

The stewardship timeline she has been on — the sequence of touchpoints, the timing of the upgrade ask, the planned giving conversation that was recently initiated — was recommended by an AI system based on patterns derived from donors who look, on paper, like her.

At what point does personalization become performance?

The answer is not determined by the sophistication of the AI. It is determined by the genuineness of the relationship. A communication that was AI-drafted but is sent by a gift officer who has actually met the donor, who knows her priorities, who has modified the AI draft to reflect something specific to their relationship — that communication is more personalized than one that a human wrote entirely from a template without that knowledge. The AI is a tool in service of a relationship that exists.

A communication that was AI-drafted, lightly reviewed for grammatical errors, and sent under a signature by a staff member who has never interacted with the donor — that is performance. The AI did not make it performance. The absence of a genuine relationship made it performance. The AI just made it easier to scale the performance to a larger audience.

The line is not between AI-assisted and human-authored. The line is between authentic and manufactured. AI can help you cross the former line. Only the actual relationship determines which side of the latter line you are on.

The donors who discover — and some will discover — that the deeply personal communication they received was AI-generated will have two different responses depending on what the underlying relationship actually looks like.

The donor who has a real relationship with the organization, who has met the people involved, who understands and believes in the mission, will process the AI involvement as a tool — like a well-designed form letter that happens to include their name and specific gift history.

They may even find it impressive. The donor who does not have a genuine relationship — who was identified by a model, communicated with through AI-generated sequences, and solicited without a single authentic personal interaction — will feel deceived. Because they were.

Trust in AI-assisted fundraising is not a technology question. It is a relationship question. The technology either supports the relationship or substitutes for it. The first builds trust. The second destroys it.

The Five Foundations

Earning trust in AI is not philosophical. It is a discipline, and it can be practiced deliberately by nonprofit leaders who understand what trust actually requires.

The first foundation is performance. Trust follows results, and results must be tracked honestly. The AI prospect model that identifies donors who subsequently give at elevated rates is earning trust with every confirmation. The one that produces high scores for donors who don't convert, without anyone in the organization examining why, is borrowing credibility it has not earned.

Every deployment should begin with a clear definition of what success looks like and a commitment to measuring it without the distortion of motivated reasoning.

When the system works, say so specifically. When it doesn't, say that too — and investigate why. Organizations that can speak honestly about both the performance and the limits of their AI tools are, counterintuitively, the organizations whose staff trusts those tools more.

Honesty about limitations is a trust signal. It demonstrates that the people running the system are not selling something they don't believe in.

The second foundation is transparency — not algorithmic transparency, which is often neither possible nor particularly useful, but operational transparency. People trust what they understand well enough to evaluate.

The gift officer who knows that a prospect score reflects the combination of giving history, event engagement, and philanthropic activity at peer institutions can weigh that recommendation thoughtfully. The gift officer who receives a number with no context cannot evaluate it, cannot push back intelligently when it seems wrong, and cannot build the kind of calibrated confidence in the tool that produces consistent use over time.

Explaining how the AI works — not at the level of mathematical architecture, but at the level of what inputs produced what outputs and why — is not optional. It is the work that converts skeptics into informed users.

The third foundation is accountability. Someone must be responsible for every decision that AI informs. This is not just a governance principle. It is a trust mechanism. People trust systems embedded in accountability structures because accountability structures mean that failures have owners — and owners have consequences.

The AI that operates in a space where no human is responsible for its outputs is not just a governance risk. It is a trust risk.

When something goes wrong and there is no clear answer to the question of who is responsible, the damage to confidence is not confined to the specific failure. It undermines confidence in the entire system, because the failure demonstrated that no one was actually watching.

The fourth foundation is fairness. The question of whether AI treats people equitably is not a nicety. It is foundational to the trust calculation of everyone being asked to rely on the system.

A prospect scoring model that consistently underestimates the giving potential of donors from certain demographic groups is not just making prediction errors.

It is making inequitable decisions that direct less relationship investment to some donors than others based on characteristics unrelated to their actual potential. Staff who observe this pattern — even without the statistical analysis to name it explicitly — will not trust the system.

They will route around it. And donors who eventually understand that they were scored and ranked in ways that reflected demographic bias rather than genuine assessment of their relationship to the mission will not trust the organization that deployed the system.

The fifth foundation is human control. The most trusted AI systems in practice are the ones that make visible and explicit — at every decision point — that a human being is in charge. Not because AI is unreliable, but because trust requires agency.

People trust tools they direct. They fear tools that direct them. The distinction sounds simple but shapes everything. The AI that surfaces a recommendation, presents the evidence behind it, and waits for a human decision occupies a fundamentally different psychological position than the AI that takes action and reports what it did.

The former is a powerful instrument in skilled hands. The latter is something else — autonomous, opaque, and outside the control of the person nominally responsible for the outcome.

That version of AI will not be trusted, no matter how often it is right, because trust is not only about accuracy. It is about the feeling of agency and the reality of accountability.

Building Trust in the Sector

The nonprofit sector faces the trust challenge in a particularly acute form, because the sector's entire legitimacy rests on trust. Not metaphorically. Literally.

Donors trust that their gifts will be used as promised. Clients trust that the organization will protect their information and their dignity. Staff trust that leadership will steward the mission honestly. Communities trust that institutions claiming to serve them are actually doing so.

The tax-exempt status that defines a nonprofit under Section 501(c)(3) is itself a trust relationship — the public extended it, through the tax code, in exchange for the organization's commitment to mission rather than private enrichment. Every element of the sector's operating framework is a trust relationship of one kind or another.

AI introduces new ways to breach that trust in ways the sector has not yet fully mapped.

A predictive model that systematically under-invests in cultivating donors from underrepresented communities — not through any explicit decision, but through the emergent logic of training data that reflects historical inequity — is encoding discrimination into the fundraising program.

A program efficiency model that identifies which clients are most likely to succeed and implicitly directs resources toward them — optimizing outcomes at the cost of mission — has mistaken efficiency for purpose.

A constituent database that ingests behavioral data from sources the constituents never knew were being collected has taken something without asking. In each case, the technical capability is real. The trust violation is also real.

And in an environment where the organization's entire operating legitimacy rests on trust, a discovered violation is not a PR problem. It is an existential one.

Lyman Pierce understood this at the structural level when he and Charles Ward built their campaigns in the early years of the twentieth century. The campaign thermometer on the courthouse square was not decoration.

It was a transparency mechanism — a daily public accounting of exactly how much had been raised and how much remained, visible to everyone in the community regardless of whether they had given.

The announced lead gifts from community leaders before the public campaign launched were not merely momentum-building. They were a trust signal: the people this community respects have already extended confidence, and had it honored.

Every structural element of the Pierce-Ward system was, beneath the fundraising mechanics, a trust architecture.

The campaigns worked because people trusted them. People trusted them because every element of the system was designed to earn and sustain that trust.

The principle is not different today. The mechanism is.

The Compounding Returns of Trust

There is a reason to do all of this work. Not just to avoid the failure modes — the adoption that stalls, the staff who routes around the system, the donor who discovers how her information was used and never gives again.

The reason to do the work is what becomes possible on the other side of earned trust.

An organization whose staff genuinely trusts its AI systems uses those systems consistently enough to generate the volume of AI-assisted decisions that improves outcomes. A fundraiser who trusts a prospect model visits the right people with the right information and closes more gifts — not because the model is magic, but because consistent use of a reliable tool compounds over time in ways that sporadic, skeptical use does not.

A program team that trusts an outcome prediction model designs better interventions — not because they have outsourced their judgment, but because they have data they believe in to inform it.

Trust is not just a precondition for adoption. It is a multiplier on effectiveness. The 80 percent adoption rate and the 30 percent adoption rate are not the same tool deployed differently.

They are two different tools, even if the software is identical — because the tool that is trusted and used consistently is producing better decisions at a volume that the tool that is doubted and used selectively is not.

The return on AI investment is not primarily a function of the quality of the algorithm. It is primarily a function of whether the people using it trust it enough to use it consistently and thoughtfully.

This is why the organizations that will get the most from AI in the coming decade are not the ones with the largest technology budgets or the most sophisticated models.

They are the ones that did the patient, unglamorous work of building trust — starting small, measuring honestly, explaining clearly, holding themselves accountable when the system failed, and demonstrating over time that the humans running the AI had the judgment and the integrity to be trusted with the tool.

Edison won not by proving electricity was superior to gas lighting. That argument was technically obvious to anyone who had stood in an electrically lit room. He won by demonstrating — installation by installation, customer by customer, year after year — that the system was reliable, that the company behind it was accountable, and that when something went wrong someone would come and fix it.

The first Pearl Street customers were not buying electricity. They were extending trust. And Edison understood, with everything in him, that the only way to honor that trust was to earn it again every single day.

The Trust That Makes Everything Else Possible

People like change. They do not like changing. The distance between those two statements is not a communications problem or a change management problem. It is a deeply human reality — written into 300,000 years of neural architecture — that demands respect before it will yield.

The nonprofit professionals sitting in the back of the conference room, the ones with their arms crossed and their skepticism intact, are not obstacles to be overcome. They are the people whose considered, informed, experiential trust is the only kind that actually holds under pressure.

The enthusiasts in the front rows are essential — they create the early evidence, the case studies, the proof that the tools can work. But it is the thoughtful middle that makes the transformation real, because the thoughtful middle does not extend trust to anything that hasn't earned it. And when they do extend it — slowly, conditionally, with their eyes open — it holds.

The cause of building trust in AI within the nonprofit sector is not separate from the cause of using AI to advance nonprofit missions. It is the same cause.

Every decision to be transparent about how a system works, every accountability structure that ensures someone is responsible for AI-informed decisions, every honest conversation about what the model gets right and what it gets wrong — those are not delays on the path to AI adoption. They are the path.

The tools Ward and Pierce had were a pencil, an index card, and the conviction that the cause was worth fighting for with everything available.

They built trust in their methods through demonstration, accountability, and the transparency of a thermometer on a courthouse square. Their campaigns worked because the people they asked to give trusted them.

That requirement — that the people being asked to trust must have reason to do so — has not changed in the century since.

It will not change in the century ahead.

Build the trust. The rest follows.

19 COMMUNICATION

"Machines take me by surprise with great frequency."
— Alan Turing

The Turing Test
In 1950, Alan Turing proposed a deceptively simple question. Not "can machines think?" — which he considered unanswerable and, frankly, not particularly interesting — but something more practical and more unsettling: can a human, through conversation alone, reliably tell the difference between a machine and another human?

Turing called it the Imitation Game. We call it the Turing Test. For 75 years it served as the distant horizon of artificial intelligence — the benchmark that serious AI would someday have to clear before we could say the machines had truly arrived.

Here is a question worth sitting with: Has fundraising already passed it?

Not in some future state. Not after the next wave of model improvements. Right now. Today.

The Test, Applied
Let's be precise about what the test actually asks. It doesn't ask whether the content is good, whether the relationship is real, or whether the communication is ethical. It asks one thing only: can you tell the difference?

Pick up the direct mail piece that arrived last week from a national health nonprofit. It references a specific disease. It tells a story about a patient. It closes with a handwritten-looking signature and a PS that creates urgency. Could you tell, reading it, whether a human writer crafted every word — or whether a large language model produced the first draft that a human lightly edited — or whether the entire piece was AI-generated and reviewed only for compliance? Most readers cannot. The piece either works or it doesn't. The origin is invisible.

Open the cultivation email from the university. It references your most recent gift, the program area you've supported, a specific outcome your giving produced. Did a gift officer write that? Did an AI generate it from structured donor data and a template? Did a hybrid workflow produce it in thirty seconds rather than thirty minutes? You probably cannot tell. And here is the uncomfortable truth: for a meaningful and growing percentage of the donor communication being produced in the nonprofit sector right now, the answer is that AI had a significant hand in it.

The Turing Test, applied to fundraising content, is increasingly a test that AI is passing.

But the test gets harder as the gift gets larger — and what happens at each level tells us something important about where AI belongs and where it doesn't.

Annual Giving: Already There

At the annual fund level, the Turing Test is effectively over. AI-generated direct mail copy, email appeals, social content, and digital ads are indistinguishable from human-written equivalents — not because AI writes with more heart, but because AI tests faster, segments more precisely, and personalizes at a scale no human copywriter can match alone.

A human copywriter produces one letter. An AI-assisted workflow produces thirty versions, each calibrated to a different donor profile, each tested against response data, each refined toward the version that actually works for that specific segment.

The chat interactions that appear on donation pages — answering questions in real time, gently reintroducing the giving opportunity at the end — are frequently AI-driven now. Most donors do not know. Most donors do not ask. The Turing Test is not being administered. It is being passed by default.

Ward and Pierce would have recognized this dynamic immediately. In 1905, the question wasn't whether to use a printing press or a handwritten note — it was which method moved the right person to act. A century later, the tools are different. The logic is not.

Mid-Level: The Interesting Zone

The mid-level segment — donors giving between $1,000 and $10,000 annually — is where the Turing Test gets genuinely interesting. This is the zone where the relationship is real enough to matter but the portfolio is too large for fully personal attention.

The personalized email that references a donor's 15-year giving history, connects their support to the specific program they funded, and closes with an ask calibrated to their giving trajectory — did that come from a fundraiser who knows this person, or from an AI system that processed their record and generated it in seconds? At mid-level, the answer is increasingly: the latter, reviewed by the former. The AI writes it. A human verifies it. The donor receives something that reads like personal attention — because it is, just not exclusively human personal attention.

The Turing Test at mid-level is not simply about whether the content is distinguishable. It is about whether the experience is distinguishable. That question is answered not by the sophistication of the AI but by the quality of the underlying relationship.

Major Giving: The Test Has a Clear Answer

At major gift level, the Turing Test resolves quickly and decisively. A seven-figure donor is not going to give because of an AI-generated message alone. They are going to be moved — or not moved — by a conversation with a person they trust, about a vision they believe in, at a moment in their life when giving at that level makes sense.

No language model produces that. No AI system closes that gift.

What AI does at the major giving level is make the human who has the conversation better prepared, more precisely timed, and more strategically positioned. The pre-visit brief that used to take hours to assemble takes minutes. The gift officer walks in better informed. The relationship advances faster.

The Turing Test is not the relevant test at this level. The relevant test is whether the human bringing the relationship to the conversation is actually ready to have it. AI is what makes the answer to that question more reliably yes.

Planned Giving: Where Stakes Become Existential

Planned giving is the most intimate category in fundraising. A donor is choosing to include an organization in what they leave behind. That decision involves mortality, family, legacy, and a level of trust that goes deeper than any other giving relationship. It is not a transaction. It is a covenant.

The Turing Test at planned giving level has a clear answer — the same answer the major gift test has. The conversation that leads a donor to name a nonprofit in their estate plan is not AI's conversation to have. It belongs to a human being who has earned the relationship and can sit with the weight of that conversation without rushing toward a close.

What AI does in planned giving is find the people who belong in that conversation before the conversation becomes impossible to have. The loyal annual fund donor who has given for twenty-three consecutive years, whose profile matches the documented characteristics of bequest donors, who has never been asked — AI finds that person. The model surfaces them. The gift officer makes the call. The Turing Test is not the question in planned giving. The question is whether the technology identifies the prospect in time for the human to build the relationship. That test, AI is passing.

The Question Nobody Is Asking

The Turing Test tells us whether content is distinguishable. It says nothing about whether it is good, true, or building the kind of relationship that leads somewhere real.

The risk in a post-Turing-Test fundraising environment is not that donors will discover they were communicating with AI. Many of them already suspect it. The risk is something more subtle: that the sector will optimize the performance of personalization without delivering the reality of it. That AI-generated communication will pass the Turing Test at every level except the one that matters — the test a long-term donor administers quietly, over years, of whether this organization actually knows them.

This is Pierce's staff-ization warning updated for the AI era. Pierce warned that organizations using paid staff to do everything that volunteers should do lose the authenticity and peer influence that make fundraising work. The same principle applies here: organizations using AI to do everything that human relationships should do will discover the same failure at greater scale and with higher production values.

The line is not between AI-assisted and human-authored. The line is between authentic and manufactured. A communication that was AI-drafted and reviewed by someone who actually knows the donor — who modified the draft to reflect the real relationship — is more personalized than one written entirely by a human from a template with no specific knowledge of the person receiving it. The authenticity comes from the relationship, not the drafting method.

A communication that was AI-drafted, reviewed for typos, and sent under a signature by a staff member who has never interacted with the donor is performance. The AI did not cause that failure. The absence of a genuine relationship caused it.

The AI just made it easier to scale the failure to a larger audience.

That distinction established, the question becomes practical: what does AI-assisted communication that actually works look like across every channel?

The Architecture Problem

Before the channel-by-channel answer, it is worth naming what most organizations are actually dealing with: a fragmented architecture that produces a fragmented experience.

The direct mail team is optimizing within its own lane. The email team is doing the same. The digital team manages its own calendar. The major gifts team runs its own cultivation sequences. Each team is working hard and acting with good intentions. The organization is producing multiple simultaneous versions of itself and sending them all to the same people.

The donor who opens an email appeal on Tuesday, receives a direct mail piece on Wednesday that quotes a different ask amount, and gets a phone call from a gift officer on Friday who has no idea either communication went out is not experiencing a communications program. She is experiencing a committee.

AI can address the architectural failure in ways that were previously impossible — coordinating across channels, suppressing conflicting communications, creating a unified view of what each donor has received and how they have responded. Shaw's observation, made more than a century ago, has never been more relevant: the single biggest problem in communication is the illusion that it has taken place. Organizations are producing more communication than at any point in their history. And first-year donor retention still hovers around thirty percent. The illusion has taken place. At scale.

Email: More Than Drafts

The first place most nonprofit communicators encounter AI is in email drafting, and many of them stop there. That is like buying a grand piano and using it as a bookshelf.

AI can draft a cultivation email in thirty seconds. That is useful, but it is not the interesting part. The interesting part is what happens when you connect that drafting capability to actual donor data.

Consider what a development professional at a mid-size environmental organization was able to build: a system that generates personalized cultivation emails for donors whose giving has shown decline signals, calibrated to their specific giving history, program interests as recorded in the CRM, and the time of year.

A donor who has given to the organization's river restoration work for six consecutive years gets a different email than a new donor who came in through the annual appeal. Not a different template with a name swapped in — a different story, told from a different angle, with different program outcomes referenced. The system produces first drafts. A staff member reviews and refines them. The emails go out in a fraction of the time they previously required.

The equity argument for this capability matters and is too rarely made. The mid-level donor segment has historically been under-served not because anyone chose to neglect them, but because the math of staff capacity didn't support personalized attention at that scale.

AI changes that math. The donor who gives $2,500 a year can now receive communication that acknowledges her history, references the program she cares about, and varies by season and context — without requiring a gift officer to manage her relationship manually.

The efficiency trap is real here and worth naming directly: the system produces the draft, and a staff member who actually knows the donor improves it. That sequence — AI as starting point, human as finishing touch — is the one that holds. The alternative, where production speed replaces human review, produces communication that performs personalization without delivering it. Experienced donors feel the difference even when they can't name it.

Direct Mail: The Oldest Channel Gets Smarter

Direct mail has been declared dead so many times that its continued effectiveness should be a lesson in epistemic humility. It still works. What it has always lacked is the kind of real-time responsiveness that digital channels allow.

AI changes that calculus. The most immediate application is segmentation — incorporating behavioral signals beyond RFM that traditional mail programs have never been able to use. A donor who has opened six consecutive emails and attended two events this year but hasn't given since last fall is a different proposition than a donor with the same giving history who has been entirely unresponsive to digital touches. The former wants engagement. The latter may need a different channel. AI can tell you which is which.

AI can also test direct mail creative at a scale previously impossible without large budgets. The healthcare organization that tested nine versions of its annual fund appeal letter using AI-assisted generation and modeling found that the version that performed best in every segment was not the version the communications team had assumed was strongest. The assumption was expensive. The test was not.

Digital and Social: The Volume Problem

Ask any nonprofit digital team what their biggest constraint is, and they will tell you the same thing: content. There is never enough of it.

A single impact story about a family whose housing was stabilized through the organization's programs can be repurposed by AI into a Facebook post, an Instagram caption, a newsletter paragraph, a mid-letter insert for direct mail, a donor impact report section, and a website update — each adapted to the tone, format, and character limits of the channel, each reviewed by a staff member who confirms the adaptation is accurate. What took a week to produce and distribute can be done in a day.

Acknowledgments: The Communication Nobody Thinks About

There is a piece of communication that most organizations send to every donor, process as quickly as possible, and think about as little as possible: the gift acknowledgment.

This is a mistake of the first order. Research on donor retention consistently identifies the acknowledgment — its timeliness, its warmth, its specificity — as one of the highest-leverage stewardship interventions available. A donor who receives a generic acknowledgment three weeks after making a gift has already moved on mentally. A donor who receives a warm, specific acknowledgment within 48 hours that references their program interest and connects their gift to a concrete outcome is in a different relationship with the organization entirely.

AI can produce personalized acknowledgments at a scale and speed previously impossible for most organizations. A healthcare foundation processing two hundred gifts a day during a year-end campaign can generate acknowledgments that reference each donor's history, acknowledge multi-year relationships, note the specific fund they supported, and connect their gift to outcomes in that program area — all reviewed by a staff member, all sent within 48 hours.

The organization that treats acknowledgments as administrative processing rather than relationship communication is making an expensive choice. It just doesn't show up immediately on any report.

Impact Reporting: Telling the Story to the Right Person
Annual impact reports are often the most expensive piece of collateral an organization produces. They are beautiful. They contain compelling stories and real data. And they are sent to every donor in the database regardless of what those donors actually care about.

AI enables segmented versions of impact reporting — not unique reports for every donor, but segmented editions that emphasize different programs, outcomes, and stories based on giving history and program interest. The organization that produces three versions of its impact report rather than one, calibrated to its major program areas, is communicating with its donors rather than to them.

More significantly, AI enables impact reporting at the individual gift level in ways previously possible only for major donors. The mid-level donor who gave $1,500 to the emergency food program can receive a brief, personalized impact statement — not a full report, but a note that says specifically what her gift made possible. The technology to produce this at scale exists. Most organizations have not yet applied it.

Internal Communication: The Overlooked Opportunity

The external communication applications get most of the attention. The internal ones may ultimately matter more. Boards of nonprofit organizations are chronically under-informed — not because staff aren't working hard, but because the preparation of board materials is genuinely time-consuming and the resulting materials rarely provide the analytical context boards need to make good decisions. AI tools can generate board-ready financial summaries and fundraising dashboards directly from underlying data, with variance analysis and narrative commentary included. The finance director reviews and presents rather than spending the week preparing to present.

Grant writing is one of the most skilled and time-consuming functions in nonprofit development. AI does not write grants. But it drafts background sections from existing materials, pulls program outcome data into proposal formats, generates first drafts of logic models, and compares drafts against RFP evaluation criteria to flag gaps before submission. None of these functions replaces the grant writer. All of them reduce the time a skilled grant writer spends on work that doesn't require her specific expertise.

Volunteer communication follows the same logic. The volunteer who doesn't receive clear instructions before an event, doesn't hear from the organization after they serve, and receives the same generic newsletter as everyone else is not being communicated with — they are being processed. AI can generate personalized pre-event and post-event communication at scale, calibrated to the specific role and the specific contribution. For organizations running twenty events a year with a hundred volunteers each, that is the difference between a volunteer program that retains its people and one that constantly recruits to backfill attrition.

The Standard That Hasn't Changed

Turing was right that "can machines think?" is unanswerable. But the more useful question — can AI help fundraisers communicate, identify, prepare, and engage in ways that serve donors and advance the mission — is answerable. The answer, across annual giving, mid-level, major gifts, planned giving, and campaigns, is already yes.

The Turing Test has been passed. The harder test begins now. That harder test is not a technology test. It is a relationships test — administered one donor at a time, over years, by organizations that have to decide whether AI is a tool in service of genuine connection or a substitute for it. The benchmark is not whether the organization is using the tools. It is whether the tools are producing better outcomes: more donors who feel genuinely known, who renew their giving, who upgrade over time, and who eventually consider the organization one they want to remember in their estate plans.

The organizations that build this right will eventually have something the sector has rarely achieved: a communication architecture that treats every donor — at every giving level — as someone worth knowing. The first-time $50 donor who receives an acknowledgment that references the specific program she supports. The five-year annual fund supporter who gets a mid-year update connected to outcomes she cares about. The mid-level donor whose cultivation emails reference her actual history. The major gift prospect who receives communication that reflects everything her gift officer knows about her, drafted in a fraction of the time and reviewed with full attention to the relationship.

This is not a technological aspiration. It is a relational one. The technology is finally capable of helping organizations live up to the relational standard their donors have always deserved.

Ward and Pierce raised $350 million with index cards and typewriters because they understood that the relationship between a donor and a cause, when it is genuine, is one of the most powerful forces in human experience. They did not win by out-communicating their contemporaries. They won by out-relating them — by building structures that put the right people in conversation with each other at the right moment, with a clear case and a compelling ask.

AI can do what they did, at scale, with a precision they never had. What it cannot do is replace the conviction they brought to every campaign — the belief that the cause was worth fighting for, the mission was worth funding, and the donor in front of them was worth knowing.

That conviction cannot be generated. It cannot be modeled. It cannot be imitated.

Everything else can. And increasingly, it is.

20 PEOPLE

"We can get used to anything — that's what makes us remarkable. And that's also what makes us responsible."
— MacKenzie Scott

The Human Element
There is a particular kind of meeting that happens in nonprofits right now, in conference rooms and on video calls and in whispered hallway conversations after the all-staff ends. The agenda item is AI. The real subject is something harder to name.

It sounds like this: *Are we going to be replaced?* Or, softer but no less anxious: *Is this still a place where what I bring to the work matters?*

These are not questions about software. They are questions about purpose. About whether the work that called people into the nonprofit sector — the human work of serving, connecting, advocating, caring — is still the core of what the organization does, or whether it is quietly becoming the decoration on top of an automated engine.

This chapter is the honest answer to that question. And the honest answer is: it depends entirely on what nonprofit leaders decide to do next.

The Brain and the Heart
There is a useful distinction to make at the outset, because the conversation about AI and nonprofit staff tends to collapse two different things into one.

AI can be the brain of a nonprofit operation. It can process data at scales no human can match, identify patterns in giving behavior that would take a team of analysts months to surface, generate draft communications faster than any individual writer, monitor hundreds of concurrent donor relationships for signals that deserve attention, flag clients who need intervention before the crisis is visible.

These are real capabilities. They are already deployed in organizations across the sector, and the results are measurable. The brain functions are genuinely AI's domain.

But the heart of a nonprofit — the reason donors give, the reason staff stay despite salaries that rarely match their talent, the reason communities trust certain organizations with their most pressing problems — that is not a function of data processing. It is a function of human presence.

The program officer who sits with a family in crisis and finds a way forward. The gift officer who remembers, without checking the database, that a donor's daughter just started college and asks how she's doing. The volunteer coordinator who calls a new volunteer on a Wednesday just to say thank you. The executive director who walks into a board meeting carrying the weight of the work and the conviction that it matters.

No model produces that. No training data captures it. No algorithm schedules it into a workflow.

Pierce understood this in ways that should still command attention. What he called "complete staff-ization" — the organizational instinct to pay professionals to do everything — was, in his view, one of the deepest threats to effective fundraising. Not because professionals are incapable. Because authenticity cannot be hired.

The volunteer who asks her peer for a gift because she believes in the cause is more persuasive than the most polished development professional precisely because her conviction is not transactional. It lives in her, not in her job description.

Substitute AI for staff-ization and Pierce's warning becomes the central challenge of this moment: the nonprofit that uses AI to replace the human elements that made it effective has not upgraded its operations. It has hollowed out its core.

What the Anxiety Is Really About

Chapter Three took staff anxiety seriously as a starting point. This chapter takes it seriously as a conclusion — the place where strategy and values and organizational design all come home.

The anxiety is not about algorithms. It is about worth. People who chose nonprofit work chose it, in most cases, for reasons that have nothing to do with processing efficiency. They chose it because they believed the mission mattered. Because they wanted to spend their working lives in proximity to something meaningful.

Because the relationship with a donor, or a client, or a community felt like the actual substance of the work — not a means to an operational end.

When those people hear that AI can now draft the appeal letter, score the prospect pool, generate the impact report, create the cultivation timeline, and automate the cultivation sequence — they are not wrong to feel that something is at stake.

They are conducting a threat assessment. The question is whether the threat they are assessing is the right one.

The real threat is not that AI will make nonprofit professionals obsolete. The real threat is that organizations will make decisions — about hiring, about job design, about what skills they invest in — that treat the human beings on their staff as a cost problem rather than as the irreplaceable source of mission delivery. The technology doesn't create that risk. Leadership choices create it.

MacKenzie Scott, whose philanthropy has moved billions of dollars to nonprofits that do hard work in communities that larger funders often overlook, has said something worth sitting with: we can get used to anything.

She means it as a warning. The same adaptive capacity that makes human beings remarkable is what makes inattention dangerous. We normalize things that should disturb us. We stop noticing what we have accepted.

The normalization risk in the AI moment is that organizations get used to the efficiency of automation and gradually stop noticing the cost. The donor retention rate that drifts down because the personal touches have been optimized away.

The staff culture that changes when people feel like monitors of automated systems rather than authors of human relationships. The community trust that erodes when clients start to feel processed rather than served.

These are not dramatic failures. They are gradual ones. They are the kind that look fine in the quarterly report and register only when you notice that something essential has gone quiet.

The Skills That Actually Matter Now

The nonprofit sector has always had a complicated relationship with professional development. Training budgets are among the first things cut when resources get tight. And the conversation about AI skills often makes the problem worse before it gets better — generating anxiety about technical capabilities that most nonprofit professionals don't have and aren't sure they need.

Here is what that conversation gets wrong: the skills that make nonprofit professionals effective in an AI-integrated environment are not primarily technical. They are primarily human — sharpened, developed, and directed with more clarity and intention than before.

Judgment is the most valuable skill in an AI-assisted organization. AI systems produce outputs. Those outputs can be impressive. They can also be subtly wrong — biased in ways that reflect historical patterns rather than current priorities, confident about conclusions that deserve more scrutiny, efficient in directions that undermine rather than advance the mission.

The professional who can evaluate AI outputs critically — who asks what the model is learning from, who notices when a prospect list skews in ways that deserve examination, who recognizes when a generated donor communication would read as authentic and when it would ring false — that professional is more valuable than ever. Not less.

Relationship fluency is the second skill. This is not a new skill for nonprofit professionals. It is the reason most of them are in the sector.

What changes with AI integration is that relationship fluency becomes the primary differentiator rather than one quality among many. The development professional who can move through AI-assisted prospecting and AI-drafted cultivation sequences and still show up to the donor conversation as a genuine human being who is actually interested in the person across the table — that professional is not replaceable.

The one who treats the AI briefing as a script and the donor conversation as a performance review of the algorithm is. This distinction matters for hiring, for training, and for how organizations define success in frontline roles.

Data literacy is the third skill — and here the sector must be honest about its deficits. The development professional who cannot read a model's outputs critically, who does not understand what a retention rate or a response rate or a propensity score actually represents, who cannot interrogate a vendor's claims about predictive accuracy — that professional is at risk of being managed by tools rather than managing them.

Data literacy does not require technical depth. It requires the ability to ask the right questions. What is this model trained on? What does this score actually predict? What are the confidence intervals? What does this output not tell me? These are not programmer questions. They are professional questions, and nonprofit organizations need to invest in building the capacity to ask them.

Narrative intelligence is the fourth. AI can generate content. It cannot generate meaning. The development professional who can take AI-assisted research and prospect analysis and construct a compelling, authentic story about why this donor's investment in this organization matters — told in that donor's language, connected to that donor's values, grounded in the specific outcomes the organization has produced — that professional brings something no language model replicates.

The story is not the output. The story is the relationship made visible. AI can scaffold it. Only a human being can make it real.

Designing Work That Doesn't Waste People

There is a version of AI integration that makes organizational sense on a spreadsheet and organizational destruction in practice. It looks like this: identify all the tasks currently performed by staff that AI can replicate, automate those tasks, reduce headcount proportionally, report efficiency gains to the board. This version is genuinely tempting when budgets are tight. It is also, in most cases, a way of destroying the organizational capacity that took years to build.

The problem is not that some tasks should be automated — they should. Grant report drafts that take twelve hours should take two. Prospect research briefings that require half a day should require an hour. Donor acknowledgment letters that have lived in a template for years should be replaced with AI-generated drafts that are actually individualized. These are legitimate efficiency gains, and the time they free should be real.

The problem is what happens to that time. If efficiency gains translate directly into headcount reductions, the organization hasn't freed up human capacity. It has eliminated it.

The development associate who no longer spends three hours a week on data entry does not simply disappear, freeing budget. She applies those three hours to work that requires human judgment — to cultivation conversations, to stewardship touchpoints, to the relationship maintenance that predicts donor retention but never appears on the efficiency report.

That is the exchange that makes AI integration valuable. That is the version that serves the mission.

Job design in the AI era requires nonprofit leaders to ask a question that most job descriptions have never been forced to answer clearly: what parts of this role are genuinely irreplaceable by automated systems, and are we protecting those parts, or are we automating them away while keeping the shell of the title?

The gift officer who spends sixty percent of her time on data entry and call logging and administrative coordination is not primarily a gift officer. She is an administrative coordinator who occasionally meets with donors.

AI should change that ratio dramatically — and the design question is whether the freed capacity is reinvested in more donor relationships or absorbed into organizational overhead. The answer to that design question determines whether AI integration serves the mission or merely reduces the payroll.

The Talent Question the Sector Cannot Avoid

There is a harder version of the people conversation, and it is about the talent pipeline.

The nonprofit sector has always competed for analytically capable professionals against private sector employers who can offer salaries the sector cannot match. This has always been a structural disadvantage. AI does not fix it. In some ways, AI makes it more acute — because the skills required to work effectively alongside AI systems in a development office are increasingly the same skills that financial services firms, technology companies, and consulting practices are recruiting for aggressively.

The data analyst who can build and interpret predictive models is not choosing between a nonprofit and a lesser option. She is choosing between a nonprofit and a fintech company that will pay her twice as much.

The communications professional who can deploy generative AI tools with strategic discipline and authentic voice is being recruited by organizations with marketing budgets that dwarf most nonprofit annual fund operations. The IT professional who understands data infrastructure well enough to govern AI vendor relationships and data security obligations is not in a talent market where nonprofits have structural advantage.

This is not an argument for despair. It is an argument for strategic clarity. The sector has always attracted talent it could not afford on salary grounds by offering something the private sector cannot: the conviction that the work matters beyond the transaction. That argument does not disappear in the AI era.

But it becomes less compelling when the work has been designed in ways that reduce the human to a monitor of automated outputs. People who believe in mission do not need to feel heroic every day. They do need to feel essential.

When AI integration is designed well, it makes mission-driven professionals more essential — not less — by giving them better information, more time, and greater capacity to do the irreplaceable work. When it is designed poorly, it makes them feel like they are waiting to be automated away. The talent consequences of the difference are not subtle.

The organizations that will attract and retain the talent they need in this environment are the ones that can make an honest case: the work we ask you to do is meaningful, the tools we give you make you more effective, and the skills you develop here will make you better at this work than anywhere else. That case is available to nonprofit leaders. It is not made by default.

The Human Choice at the Center

Every other chapter in this book has argued, from different angles, that AI is a tool — one that can be deployed wisely or poorly, in service of mission or in displacement of it, with human judgment at the center or without it.

This chapter makes the same argument from the inside: from the perspective of the people who come to work every day, carry the mission in their professional identities, and are trying to figure out where they stand in an organization that is changing faster than most of them anticipated.

Those people deserve a clear answer from their leaders. Not reassurance that nothing will change — that is not true and the people hearing it know it isn't. Not anxiety-amplifying silence about decisions that are already being made — that destroys exactly the trust on which effective organizations run. What they deserve is an honest accounting: here is what is changing, here is why, here is what we value in the people who work here that no technology replaces, and here is how we are designing this organization to protect and develop that value.

MacKenzie Scott's words carry weight precisely because she has chosen to place her philanthropy where other funders have not — in organizations that do essential work without the institutional prestige or the name recognition that attracts easy money.

She has bet, repeatedly and at extraordinary scale, on the proposition that the people doing the work matter more than the brand doing the work. That is a conviction about human capacity, about the irreducible value of people who are proximate to problems and trusted by the communities they serve.

That conviction is the right one for this moment. Not because it is comfortable. Because it is accurate.

AI will get better. The models will improve, the costs will drop, the capabilities will expand in ways that are not fully predictable today. The pace of change is not slowing.

What will not change is the fundamental truth that Pierce identified a century ago and that has survived every technological transition since: the cause is larger than the tool, the relationship is more powerful than the process, and the human being who believes in the work and shows up for it — with full presence, genuine conviction, and the skills to deploy every available resource in service of the mission — is still, and will remain, the irreplaceable core of everything a nonprofit does.

AI might be the brain.

People are the heart.

The heart is not optional.

21 FUTURE

"The future is not created by chance. It is created by choice."
— Jim Rohn

Unintended Connections

Basile Bouchon was not thinking about AI. He was thinking about silk.

The son of an organ maker, Bouchon applied the logic of his father's trade to a loom — replacing the mechanical cylinder with a roll of perforated paper. The pattern existed before the weaving began. That is the insight worth carrying forward.

Three hundred years later, every database, every predictive model, every AI system running in every development office on earth is downstream of that moment in Lyon. The principle hasn't changed. The scale has become almost incomprehensible.

The Long Line

From Bouchon's workshop, the logic of encoded instruction traveled through Jacquard's loom of 1804, through Babbage's analytical engine, through Hollerith's tabulating machines that processed the 1890 U.S. Census in weeks rather than years.

Then 1905, and Ward and Pierce running campaign accountability structures on index cards and telegraph wires, raising more than $350 million before 1923. Then 1979, and Anthony Bakker winning a contract from the Nightingale-Bamford School and calling his company Blackbaud. Then 2016, and the argument in *Data Driven Nonprofits* that the sector was sitting on hidden treasure it didn't know how to use. Then now.

Each generation inherited the failures and partial successes of the one before. Each generation punched new holes in the paper, gave the machine a little more to work with, and got a little closer to something that looked like genuine intelligence. The tools will keep changing. That part of this story is not close to finished.

What Is Actually Coming

Most books about AI play it safe in the future chapter. Careful hedges. Vague projections. Responsible uncertainty. That approach is honest — no one predicted in 1725 that perforated paper would lead to language models capable of drafting a bequest proposal. Specific predictions about a technology moving this fast are exercises in humility, not foresight.

But pattern recognition is different from prediction. And the patterns here are clear enough to say some things plainly. AI is not going to remain a workplace tool. It is already something larger than that — and the nonprofit professionals reading this book will feel its effects far beyond the development office, in ways that will reshape what the sector is being asked to do and who it is being asked to serve.

Consider what is already happening in medicine. AI systems are reading pathology slides and identifying cancers that trained radiologists miss. They are predicting sepsis onset hours before clinical symptoms appear. They are flagging drug interactions that pharmacists catch only when a patient is already at risk. A 2023 study in *Nature Medicine* found that an AI diagnostic system matched or outperformed dermatologists in identifying melanoma from clinical images. The technology is not replacing physicians. It is giving them capabilities that no individual physician, however skilled, could develop through clinical experience alone.

Earlier detection. More accurate diagnosis. Better outcomes. The patients who benefit from this are the same people whose health drives their philanthropic capacity, their engagement with mission, their decisions about estate planning. The AI that helps a donor's oncologist catch a recurrence earlier is not separate from the AI that helps a gift officer prepare a better visit. It is the same technological wave, arriving at different shores.

The same pattern is playing out in education. AI tutoring systems are providing personalized instruction at a scale and patience no human teacher can match — adapting difficulty in real time, identifying misconceptions before they calcify, working with students at two in the morning when no teacher is available.

Early evidence from programs in South Asia and sub-Saharan Africa suggests measurable improvements in foundational literacy and numeracy. The students being reached are the future donors, volunteers, and advocates for the organizations doing this work. Their education is not incidental to the nonprofit sector's future. It is constitutive of it.

In mental health — one of the most resource-constrained fields in human services — AI-powered tools are extending the reach of an entirely insufficient clinical workforce. Crisis text lines use AI triage to identify the conversations most urgently requiring a human counselor.

Therapy support apps provide cognitive behavioral interventions between clinical appointments, reducing symptom severity in studies of anxiety and depression. These tools are not replacing therapists. They are doing what therapists cannot do: being available at three in the morning to someone who is not safe.

In workforce development, AI is personalizing job matching, skills gap analysis, and career pathway mapping in ways that case managers managing forty clients at a time cannot replicate manually.

In housing stability programs, predictive models are identifying which families are most likely to experience eviction six months before it happens, enabling intervention while options still exist.

In food security, AI-optimized distribution logistics are reducing waste and improving delivery to the households that need it most.

This is the reality that nonprofit leaders need to hold alongside everything else this book has argued: AI is not just changing how organizations operate. It is changing the problems those organizations are being asked to solve, the populations they are being asked to serve, and the expectations — from funders, from government, from communities — about what effective service delivery looks like.

The health organization that deployed AI to improve grateful patient identification is now being asked by its hospital partners to help think about AI-assisted patient navigation.

The workforce development program that built a predictive model for employment outcomes is now being evaluated by foundations that expect to see that level of analytical rigor across the portfolio. The standard is moving. It will not stop moving.

The Personal Dimension

There is a version of this conversation that stays comfortably abstract — sector-level trends, organizational capability, strategic positioning. It misses the most important thing.

AI is going to change what it means to be a person in the world over the next twenty years in ways that dwarf its impact on any particular professional practice. The nonprofit professionals reading this book will experience this personally, not just organizationally.

Their parents will interact with AI-assisted healthcare systems that catch things earlier and communicate more clearly. Their children will be educated in environments where AI tutoring is as normal as a textbook. Their own experience of navigating insurance, legal systems, financial planning, and public services will be shaped by AI tools they interact with daily — some well-designed and genuinely helpful, others producing the nonprofit sector's equivalent of the vodka-and-meat translation problem with bureaucratic confidence.

These personal experiences will shape what the people who work in nonprofits believe about AI. The gift officer whose mother received an AI-assisted cancer diagnosis six months earlier than she otherwise would have is not going to approach AI skepticism the way she might have otherwise. The program director whose daughter learned to read with the help of an adaptive AI tutoring system is not going to have theoretical views about what AI can and cannot do for the people his organization serves.

The development director who spent hours navigating a broken AI chatbot trying to resolve an insurance claim has a data point about what happens when the tool is deployed without adequate human oversight.

Experience is the fastest trust-builder and the fastest trust-destroyer. As AI becomes woven into the texture of daily life — in healthcare, in education, in financial services, in government — the nonprofit sector's staff, donors, volunteers, and board members will arrive with accumulated experience that shapes every conversation about AI adoption. Some of that experience will be with AI that worked well and made things genuinely better.

Some of it will be with AI that failed in ways that were costly, opaque, and hard to appeal. Both kinds of experience will be present in every room where the decision to deploy a new AI tool is being discussed.

The sector that understands this — that takes the personal dimension of the AI era seriously rather than treating it as background noise — will have a different kind of conversation with its stakeholders.

Not just about what AI can do for the organization, but about what people have already experienced with AI elsewhere, what they trust and what they don't, and what it would take to build the specific kind of trust that this organization, with its specific mission and constituent relationships, actually needs.

Pierce built his campaigns on the conviction that the cause was worth fighting for. That conviction was personal before it was professional. The nonprofit sector's relationship with AI will ultimately be the same: personal before it is institutional, experiential before it is strategic, and determined more by the accumulated daily encounters of millions of individuals than by the policy decisions of any organization or government.

The Choices That Create the Future

Jim Rohn was right about something that gets paraphrased often enough to lose its force: the future is not something that happens to us. It is something we construct, decision by decision, in the present.

Data Driven Nonprofits made a version of this argument in 2016. The data was always there. The question was always what organizations chose to do with it. *How to Raise Money* makes the same argument across a century of evidence.

The principles that Pierce established in 1905 are recognizable in every effective development program running today — not because the world hasn't changed, but because human motivation hasn't changed, and the patterns of giving behavior that emerged from thousands of campaigns encoded truths about human psychology that no technological progress has overwritten.

AI Driven Nonprofits is the third point in this line, and the argument running from the beginning to the end of this book is the same argument: the future of the nonprofit sector is not being written by the technology. It is being written by the choices that leaders make about how to use it.

Those choices have a pattern too.

The organizations that will thrive in the AI era are the ones that made a prior choice to be data driven — that did the hard, unglamorous work of data quality, built a culture of evidence, and learned to measure what matters rather than what is easy to count. For them, AI is not a revolution. It is a natural continuation.

Data gave them eyes to see patterns. AI gives them the capacity to act on those patterns at a scale no previous generation of nonprofit leaders has had access to. The organizations that tried to skip the data step and jump directly to AI will find themselves running a Ferrari without fuel — impressive in the driveway, useless on the road.

The organizations that chose the efficiency trap — that used AI to eliminate the human elements of fundraising rather than augment them — will discover, with a lag that makes the cause hard to trace, that their retention rates have drifted, their major gift pipelines have thinned, and their donor relationships have the texture of transactions rather than commitments.

Pierce warned about complete "staff-ization" — paying professionals to do everything, assuming authenticity could be hired. The modern version of that mistake is complete automation: assuming that optimizing the process preserves the relationship. It doesn't.

And the organizations that led with the human element — that used AI to find the people, and then deployed real human beings to sit across from them and ask — those organizations are building something Pierce would recognize. The tool has changed. The fundamental act has not.

What We Cannot Predict, and What We Can

No one predicted in 1725 that perforated paper would lead to systems that can analyze every interaction a donor has ever had with an organization and surface the three people most likely to make a transformational gift in the next eighteen months.

No one in 1905 predicted that the accountability structures Pierce built into his campaigns would find their digital expression in CRM dashboards that his gift officers could operate from a mobile phone.

We cannot predict, in any reliable detail, what the tools will look like in 2050. The pace of change is not linear. Capabilities that seemed years away are arriving in months. The honest position is that the specific landscape is not knowable from here.

But the patterns are knowable. They have always been knowable.

People will still give to causes they care about, presented by people they trust. Major gifts will still require personal relationships developed over time. The specific ask will still outperform the vague request. The lead gift will still determine campaign success. Stewardship will still determine retention.

And the human being who believes in the mission and shows up for it — with full presence, genuine conviction, and the skills to deploy every available resource in service of what the organization exists to do — will still be, as Pierce understood a century ago, the irreplaceable core.

AI will get better at finding those people. It will get better at understanding their patterns, predicting their readiness, and personalizing the approach. It might, in time, handle functions we cannot yet imagine automating.

But it will not replace the thing Pierce spent his career building: the conviction-driven ask. The moment when one human being turns to another and says, simply, *this matters, and I believe you are someone who understands that.*

That is not a data point. It is not a pattern. It is not something that can be learned from training data or optimized by a model. It is the beating heart of this entire enterprise — the irreducible human center around which every tool, from the perforated paper to the punch card to the predictive model, has always been arranged.

The Choice in Front of You

Every chapter in this book has been, at its core, about a choice. The choice to invest in data quality before claiming AI readiness. The choice to define the problem before selecting the tool. The choice to maintain the human element rather than optimize it away. The choice to hold the accountability structure that Pierce built into every campaign — because accountability is not a reporting function, it is the structure that changes behavior. The choice to see AI as a tool in service of mission rather than a shortcut around the hard work that mission requires.

Those choices are not made once. They are made in every budget cycle, every hiring decision, every conversation about how to deploy a new capability. They are made when someone asks whether this AI tool will let the organization reduce headcount — and the answer to that question is itself a choice about what kind of organization is being built. They are made when the board asks whether AI is producing results, and the answer requires being honest about whether the organization measured outcomes or merely adoption.

Lyman Pierce finished his manuscript in San Francisco in 1932, thirty years after he began his career in fundraising, and opened with a simple statement: he had learned much of what *not* to do.

He wrote the book because he believed the accumulated wisdom of those decades deserved to be passed forward — because he could see, even then, that the principles his campaigns had validated were not temporary adaptations to a particular moment. They were patterns. Durable, proven patterns about how human beings relate to causes they care about, to organizations they trust, and to the act of giving itself. He was right. The patterns held across ninety years and two technological revolutions before this book carried them forward again.

The patterns will hold across whatever comes next. The tools will change. The models will improve. The capabilities will expand in ways that would have seemed like science fiction at every prior point in this long story. The world outside the development office will keep changing too — in hospitals and classrooms and housing agencies and workforce centers — in ways that will reshape what the nonprofit sector is asked to do and who it is asked to serve.

None of that changes what Pierce knew and Ward stood beside him to validate: that organized information, disciplined process, and people who believe in what they are doing are still the irreducible combination. The AI now available to the sector is, for the first time, an instrument powerful enough to honor those principles at a scale Pierce could not have imagined.

The pattern has been running for three hundred years. The loom is better than it has ever been. The question, as it has always been, is whether you know what you want to weave. The choice is yours. It always has been.

It worked in 1725. It worked in 1905. It worked yesterday. The patterns are all around us. The tools have never been better. The principles have never changed.

EPILOGUE: ASK PIERCE

A Conversation with 100 Years of Fundraising Intelligence
There is a certain irony in the fact that the man who invented modern fundraising never owned a computer, never sent an email, and never once ran a predictive model. Lyman Pierce built the architecture of the entire profession — prospect lists, campaign structures, gift tables, volunteer hierarchies, public scoreboards, accountability metrics — using a card file, a telegraph, and an unshakeable conviction that institutions worth funding were worth fighting for.

He would find the current moment fascinating.

What follows is something different from the chapters that preceded it. Not more analysis. Not another framework to implement. This is a strategic playbook — built from the accumulated intelligence of Pierce's original 1932 work, the research and data behind *Data Driven Nonprofits*, and the principles woven through every chapter of this book — synthesized into the kind of direct, sequential counsel Pierce himself would have offered a nonprofit leader sitting across from him in his San Francisco office.

Call it Pierce. An agent that does exactly what Pierce did in 1932: looks at your situation plainly, draws on a century of proven experience, and tells you what to do.

The questions that follow are the ones every serious nonprofit leader should be asking. The answers are drawn from the best of what we know — not what is fashionable, not what is being sold at this year's conference — but what has been proven, across a hundred years of organizational practice, to actually work.

"Where do we start?"
Pierce answered this question the same way every time. Not with technology. Not with a campaign. Not with a strategic planning retreat.

Start with your data.

This is not a modern obsession. Pierce's very first operational step in any engagement was building what he called the listing — a complete, accurate, capacity-rated record of every potential donor in the organization's universe.

He understood something that took the rest of the sector decades to rediscover: missing data is missing money. If a person capable of giving $100 is not on your list, you have already lost that $100. The absence of a record is the absence of a relationship.

Data Driven Nonprofits documented what this looks like in modern terms. Address validation, at minimum four times a year. Deduplication after every significant import. Annual demographic appends. The No Garbage Rule: collect what is actionable, clean what is valuable, and do not keep data you have no intention of using.

The organizations that successfully adopt AI in the next decade will not be the ones who move fastest. They will be the ones whose data was ready. AI learns from your history. If your history is incomplete, biased, or corrupted, your AI will be too — efficiently and at scale, with the same confident velocity it applies to everything else.

Before the CRM. Before the model. Before the AI vendor presentation. Start with your data.

"We have data. What do we do with it?"
Ask better questions.

This is the discipline that separates organizations that extract value from data and those that simply accumulate it. Pierce asked specific questions before every campaign: Who are the prospects? What are their capacities? What are the relationships between them? What is the realistic gift table?

He did not gather information in the abstract. He gathered information in service of a defined problem.

The modern equivalent is the Nonprofit Data Paradox: more data, if not managed in service of specific questions, produces less value, not more. A 360-degree view of your constituents without a direction to move in just leaves you running in circles.

The question to ask is: What is hiding in our data that we are not yet seeing?

It might be the pattern that predicts which $250 donors are six months away from their first $1,000 gift. It might be the lapse signals that appear 90 days before a longtime donor goes silent. It might be the demographic indicator that identifies planned giving prospects in a file you've been treating as purely transactional.

Every organization has its strawberry Pop-Tart — the insight that would change behavior immediately if you knew it was there. The question is whether you are curious enough to look.

Start with the right question. Then and only then, select the tool that answers it.

"Should we be using AI?"
Wrong question.

The right question is: Do we have the foundation that makes AI useful?

Pierce was direct about this. He called it mastering the fundamentals before adding sophistication. He had no patience for organizations that chased clever tactics before getting the strategy right — shiny object syndrome in any era, and what the AI moment has supercharged into an organizational risk.

Here is the honest diagnostic. If your donor retention rate is below 45%, AI will not fix it. If your data is missing email addresses for more than half your file, AI will not fill in the gaps. If your organization has never successfully acted on a predictive model, it is not ready to deploy generative or agentic AI. These are not failures — they are starting points. But they require sequencing.

The maturity scale still matters. Raw data collection and cleaning come first. Descriptive analytics — understanding what happened — come next. Diagnostic capability — surfacing why — follows. Predictive modeling builds on all of it.

And AI, in its most powerful forms, extends the scale further than any of these predecessors. But it extends it. It does not replace it.

An organization cannot skip from spreadsheets to AI agents and expect the gap to close itself.

When the foundation is solid, AI is force multiplication. When the foundation is missing, AI is expensive decoration.

"What kind of culture do we need to build?"
The culture you build before the technology arrives determines what the technology does when it gets there.

Pierce identified this principle a century ago, before the word algorithm existed. He called it burning conviction — the single most vital factor in any great undertaking. Not the sophistication of the methods. Not the size of the budget. The depth of the belief that the work matters enough to do it right.

In modern terms: leaders must ask hard questions before they deploy new tools. Staff must be analytically literate enough to evaluate AI outputs critically — not just accept them.

Champions must exist at every level of the organization, not just in the C-suite. Data must be part of the daily workflow, not an annual exercise in reporting.

Data Driven Nonprofits identified five characteristics of organizations that successfully became data-driven: natural curiosity, data health habits, champions at all levels, data embedded in daily work, and the ability to tell stories from data. These are not technical competencies. They are cultural ones. They apply equally — more urgently — to AI readiness.

Pierce also warned against what he called complete staff-ization: paying people to do everything and losing the volunteer authenticity that made fundraising work. The AI equivalent is automating everything and losing the human judgment that makes relationships real. The organizations that will look back in a decade and recognize themselves as genuinely AI-driven are the ones that kept human beings at the center — and used AI to make those human beings more informed, more timely, and more effective.

Culture is not the soft part of AI adoption. Culture is the hard part. It is also the only part that cannot be purchased.

"How do we explain the importance of AI to our board?" Start by not calling it a technology briefing.

The moment you frame the AI conversation as a technology update — another agenda item between the investment committee report and the facilities request — you have already lost. Boards are not equipped to evaluate technology. They are equipped to evaluate strategic decisions, fiduciary responsibilities, and mission alignment. That is exactly what the AI conversation is about, if you frame it correctly.

Pierce knew how to work with a leadership body. He knew that board members didn't need to understand the mechanics of a campaign to lead one. They needed to understand the stakes, the strategy, and their specific role.

The board that understood *why* the goal was $2 million, *what* would happen if the campaign fell short, and *what* was being asked of them specifically — that board led effectively. The board that got buried in campaign tactics became a committee of skeptics.

The same principle applies to AI. The board conversation has three parts, and none of them require a technical vocabulary.

The first is the strategic case. AI is not an optional upgrade for organizations that can afford to be sophisticated. It is a structural shift in what effective nonprofit operations look like — and the gap between organizations using it well and organizations ignoring it is compounding every year.

The board needs to understand this not as an abstraction but in concrete terms: What are peer organizations doing? What will it cost to close the gap if the organization waits? What specific mission outcomes are at stake?

Pierce would have called this knowing the field. The board that understands the full landscape — not just the organization's internal situation — can make a strategic decision rather than a reactive one.

The second is the fiduciary responsibility. When AI tools are processing donor data, generating constituent communications, and informing resource allocation, the board that has never asked how those tools work — or whether the organization's use of them is consistent with its obligations — is abdicating a responsibility it probably doesn't realize it has.

The leadership team that doesn't bring AI to the board is making a governance decision by omission. Neither is defensible when something goes wrong.

The third is the specific question the board actually needs to answer: Are we satisfied that our use of AI — or our decision not to use it — is consistent with our obligations to the people we serve, the donors who fund us, and the mission we exist to advance? That question doesn't require technical expertise. It requires the same judgment a board applies to every other significant organizational decision.

Prepare three things before the board conversation. A one-page summary of what the organization is currently doing with AI, what it is considering, and what it is not doing. A clear statement of the specific mission problem AI is being used to address — not "we are exploring AI" but "we are using predictive modeling to identify donors at risk of lapsing, which we expect to improve retention by X percent."

And a proposed governance framework the board can approve, revisit annually, and point to as evidence that AI deployment is being managed with intention.

The board that approves a framework and receives honest annual updates on performance against it is doing its job. The board that never sees the question is not.

"How do we explain how we are using AI to donors and funders?"
The question worth asking first is: would they feel deceived if they found out on their own?

That is the practical test — not the legal one, not the technical one, but the relational one. The nonprofit sector was built on trust. Donors trust that their gifts will be used as intended. Funders trust that the organizations they support are operating with integrity.

The moment a donor or funder discovers that AI has been woven into the organization's operations in ways they weren't told about, the response depends entirely on whether the discovery feels like a betrayal or a revelation.

Pierce was meticulous about constituent communication not because it was legally required but because he knew that earned trust was the most valuable thing a nonprofit could possess — and careless communication destroyed it faster than anything else.

The transparency obligation with donors and funders operates at three levels.

The first is honest disclosure about how constituent information is used. Donors who gave their name, address, and giving history to support a cause they believe in did not necessarily consent to becoming the subject of a detailed predictive profile that incorporates external wealth data, public records, and behavioral signals from multiple sources.

Most donors understand and accept that organizations they support will use reasonable information to communicate with them.

Most donors do not know the extent of external data being appended to their records. Closing that gap — through updated privacy practices, plain-language disclosures, and staff who can explain the process when asked — is not a burden. It is a baseline.

The second is authenticity in communication. There is a line between AI-assisted and AI-manufactured, and every donor communication either lands on one side of it or the other. A development officer who uses AI to research a donor's philanthropic history and draft the opening of a visit briefing — and then shows up to that conversation as a genuine human being who is actually interested in the person across the table — is using AI to be more effective. An organization that deploys AI-generated cultivation sequences to donors who have never had a real interaction with anyone on staff, and calls those sequences personalized engagement, is performing personalization.

Donors who discover the difference respond based on whether the underlying relationship is real. The ones with genuine relationships understand. The ones without genuine relationships feel what they are: processed.

The third is what funders specifically need to see. Foundations and major institutional funders are increasingly asking grantees to describe their AI practices as part of due diligence. The organizations prepared for that question — with a clear policy, documented use cases, governance structures, and honest measurement of outcomes — will be better positioned than those scrambling to assemble an answer after the question arrives.

The simple standard: use AI in ways that would make donors feel known and valued if they were aware of it. Disclose what you are doing in plain language. Keep humans in the loop for every consequential communication. Measure the results honestly and share them.

Trust in AI-assisted fundraising is not a technology question. It is a relationship question. The technology either supports the relationship or substitutes for it. The first builds trust. The second destroys it.

"What should we do to prepare our staff to take advantage of AI?"
Stop training for technical competency and start developing judgment.

That is the mistake nearly every organization makes when it first confronts AI staff development. The assumption is that the missing ingredient is technical knowledge — that staff need to understand how large language models work, how predictive algorithms generate scores, how to write a prompt. Some of that is useful. None of it is the point.

The skills that make nonprofit professionals effective in an AI-integrated environment are not primarily technical. They are primarily human — sharpened, developed, and directed with more clarity and intention than before.

Judgment is the first skill — and the most valuable. AI systems produce outputs. Those outputs can be impressive. They can also be subtly wrong, biased in ways that reflect historical patterns rather than current priorities, confident about conclusions that deserve more scrutiny.

The professional who can evaluate AI outputs critically — who asks what the model is learning from, who notices when a prospect list skews in ways that deserve examination, who recognizes when a generated donor communication would read as authentic and when it would ring false — that professional is more valuable in the AI era than before it. Train for the capacity to question the machine, not just to operate it.

Relationship fluency is the second skill — and here the nonprofit sector has a structural advantage it must not squander. The development professional who can move through AI-assisted prospecting and AI-drafted cultivation sequences and still show up to the donor conversation as a genuine human being who is actually interested in the person across the table is not replaceable. The one who treats the AI briefing as a script is. The training question is not how to use the briefing tool — it is how to use it and remain fully present in the room.

Data literacy is the third skill — and here the sector must be honest about its deficits. Not statistical modeling fluency. Not programming. The ability to read an AI output and understand what it is based on, what it might be missing, and when to override it.

The gift officer who looks at a prospect score and knows the model can't see the offline relationship the organization has built over five years. The program director who recognizes that a client prioritization model trained on historical service data is reflecting who has been served, not necessarily who most needs to be served. This kind of practical data literacy is teachable. It is also urgently needed.

Pierce was meticulous about volunteer and staff preparation — he believed an untrained person in the wrong role caused more damage than an absence in that role. His standard was not maximum sophistication. It was minimum competence in the specific capacities the work required. Apply the same standard to AI readiness: identify the capacities the work actually requires and build them deliberately.

In practice, this means three things. Create protected time for staff to experiment with tools before they are deployed in production — the team that has never played with AI will not use it effectively when it matters. Build human review requirements into every AI-assisted workflow from the beginning, as a design principle rather than an afterthought.

And reward the staff member who catches a model error. The organization where staff feel safe raising concerns about AI outputs learns from them. The organization where they don't amplifies its problems until they become crises.

The goal is not a staff that trusts AI. It is a staff that evaluates AI — and brings everything that makes them irreplaceable to the work that comes after the evaluation.

"How do we recruit and hire new team members that embrace the change that comes with AI?"
Rewrite the job description. Then rewrite the interview.

Most nonprofit job descriptions were written before AI was part of the operational landscape, and they describe roles that no longer exist in quite the form intended. The development associate who will spend sixty percent of her time on data entry and administrative coordination in a role designed before AI is now a different role than it was three years ago — whether the job description reflects that or not. The gap between the description and the reality is a recruiting problem before it becomes a retention problem.

The first step is honest job design. Pierce warned against paying people to do everything and losing the authenticity that made fundraising work. The modern version of that mistake is hiring people into roles defined by tasks that AI will automate within the first year of their tenure, without being transparent about what the role will actually require.

That produces staff who feel deceived — and organizations that lose good people precisely when the investment in them is paying off.

The question every job description must now answer explicitly: What parts of this role require irreplaceable human judgment, relationship, and conviction — and what parts will be increasingly assisted or automated? The honest answer to that question describes the actual role and attracts candidates suited to it.

There are two candidate profiles the sector consistently undervalues and should now actively recruit.
The first is the analytically curious generalist. Not a data scientist — a fundraiser, program officer, or communications professional who is genuinely curious about why things happen, comfortable working with information, and intellectually interested in what the data is showing.

Curiosity — specifically the impulse to ask better questions rather than accept the first answer — is the disposition that makes AI useful rather than merely operational. The candidate who, in an interview, asks what are we not seeing in our data rather than what tools do you use is demonstrating exactly the mindset the moment requires.

The second profile is the mission-committed professional with private sector analytical experience. The sector has always struggled to compete on salary for this talent. The competition is now more acute — the skills required to work effectively alongside AI systems in a development office are increasingly the same skills that financial services firms and technology companies recruit for aggressively.

The answer is not to pretend the salary gap doesn't exist. It is to make an honest and specific case for why the work itself is worth the differential: the conviction that what you are doing matters beyond the transaction, combined with genuine organizational commitment to deploy AI in ways that make the human professional more capable and more essential, not less.

That case — we will give you better tools, more meaningful work, and real investment in your growth — is available to nonprofit leaders. It is not made by default.

The organization whose AI integration has been designed to free human capacity for higher-value work can make it honestly. The organization whose AI strategy is primarily about reducing headcount cannot.

In the interview, stop asking about experience with specific tools. Tools change faster than hiring cycles. Ask instead: Tell me about a time you questioned a recommendation and why.

Describe how you make a decision when the data points one direction and your instinct points another. What does it mean to you that the work we do matters? The candidate who can answer those questions has the foundation that AI-era nonprofit work requires. The candidate who can only list software platforms has the foundation for a resume, not a role.

Pierce's standard for campaign leadership was not technical sophistication. It was burning conviction combined with systematic discipline. The combination — believing the work matters and doing it rigorously — is what he hired for, developed, and promoted.

That standard has not changed. The tools around it have.

"What is the one thing we cannot automate?"
Pierce answered this question in 1932. It has not changed.

Conviction.

The ability to look another person in the eye and say: This cause is worth your investment. I believe it. I've seen the evidence. I'm asking you directly to help.

AI can identify who that conversation should be with. It can tell you the right moment to have it, the right amount to ask for, the most relevant impact story to lead with. It can draft the follow-up, schedule the next touch, and flag the moment engagement starts to fade.

It can do all of this — and it cannot do the asking. Not really. Not in the way that moves someone to write a check that changes their life and changes the lives of people they will never meet.

Pierce raised more than $350 million in his career — over five billion in today's dollars — without a single digital tool. What he had was actual intelligence: about donors, about community, about what moves people to act.

He had data in the form of card files and personal knowledge. He had analytics in the form of pattern recognition and experienced judgment. He had the human element in the form of well-trained, motivated people who believed in the goal and never stopped asking.

The tools have changed. The intelligence required to use them wisely has not.

AI can find the people who need to hear your case. It can help you prepare to make it. It can make the relationship more informed, more timely, and more personal. It cannot make it less real. It cannot provide the fire.

That part has always been on us.

Notes

AI Driven Nonprofits completes a trilogy of books starting with *Data Driven Nonprofits* in 2016 and *How to Raise Money* in 2026. They have been professional and personal projects for me.

I want to start by thanking both Roger Craver and Beth Kanter for writing the “Foreword” and “Fast Forward” openings to the this book. Roger and Beth have been mentors and friends for many years now. Their wisdom and help have been invaluable to me in so many ways.

I need to thank Anthropic’s Claude Sonnet 4.6 model for helping to edit, re-edit, and index this book. This included developing an AI skill based on everything I’ve written over the last 10 years to help to give context and keep my voice intact.

I must thank everyone who’s attended keynote presentations, workshops, webinars, read articles, or purchased books over these many years. You have help shape my thinking and my life. Your feedback has been a gift.

Finally, thank you to my two best boys and family for their love and support on this journey we call life.

This book has been brought to you by the letters A and I.

Index

A

B

C

D

F

G

N

O

P

R

S

T

U

W

Y

www.ingramcontent.com/pod-product-compliance
Lightning Source LLC
LaVergne TN
LVHW052244200926
848481LV00002B/263
* 9 7 8 0 9 8 8 8 5 0 7 2 9 *